BIOLOGICAL NETWORK ANALYSIS

BIOLOGICAL NETWORK ANALYSIS
Trends, Approaches, Graph Theory, and Algorithms

PIETRO HIRAM GUZZI
University of Magna Græcia
Catanzaro, Italy

SWARUP ROY
Sikkim (Central) University
Sikkim, India

ELSEVIER

Elsevier
Radarweg 29, PO Box 211, 1000 AE Amsterdam, Netherlands
The Boulevard, Langford Lane, Kidlington, Oxford OX5 1GB, United Kingdom
50 Hampshire Street, 5th Floor, Cambridge, MA 02139, United States

Notices

Knowledge and best practice in this field are constantly changing. As new research and experience broaden our understanding, changes in research methods, professional practices, or medical treatment may become necessary.

Practitioners and researchers must always rely on their own experience and knowledge in evaluating and using any information, methods, compounds, or experiments described herein. In using such information or methods they should be mindful of their own safety and the safety of others, including parties for whom they have a professional responsibility.

To the fullest extent of the law, neither the Publisher nor the authors, contributors, or editors, assume any liability for any injury and/or damage to persons or property as a matter of products liability, negligence or otherwise, or from any use or operation of any methods, products, instructions, or ideas contained in the material herein.

Library of Congress Cataloging-in-Publication Data
A catalog record for this book is available from the Library of Congress

British Library Cataloguing-in-Publication Data
A catalogue record for this book is available from the British Library

ISBN: 978-0-12-819350-1

For information on all Elsevier publications
visit our website at https://www.elsevier.com/books-and-journals

Publisher: Mara Conner
Acquisitions Editor: Chris Katsaropoulos
Editorial Project Manager: Ana Claudia A. Garcia
Production Project Manager: Prasanna Kalyanaraman
Designer: Miles Hitchen

Typeset by VTeX

To Anna, Fernando, Salvatore, Fernando Jr., and Pietro

—**Pietro H. Guzzi**

To Lt. Sukumar Roy (father), Lt. Subhash C Karmakar (father-in-law), and other members of my loving family (Dida, Maa, Mother-in-law, Chumki, Arpita, Srinika, and Shaunak)

—**Swarup Roy**

Contents

Foreword

Graph or network has become a powerful representation tool to specify relationships among data, and the graph theory has established a fundamental vehicle to mathematically model the pairwise relationship among different objects or entities. Consequently, graph has been popularly used to model various data from many applications. In biology, biological systems are composed of basic building blocks made up of mutually interacting cellular components. For example, it has been observed that proteins seldom act as single isolated species in the performance of their functions; rather, proteins involved in the same cellular processes often interact with each other. Such interaction relationships can be well modeled by a graph or network structure. Generally speaking, graph-based biological networks have become a novel mathematical tool to model the complex association among various cellular components, such as genes, proteins, and metabolites. It is also important to model them mathematically and computationally to derive a sketch on the interdependent relationships among the cellular components to elucidate such unknown rhythmic activities within a cell.

This book offers a comprehensive reading on biological network analysis. It includes fundamental and in-depth introductory materials on graph and graph theory. The book then demonstrates how the graph is applied to analyze various different types of biological networks. In particular, this book focuses on three types of biological networks: gene expression network, protein-protein interaction network, and brain connectome network. In-depth details are provided in separate chapters for each kind of the network, their properties, the process of generating them, and recent trends and tools in analyzing those networks. The discussion can be easily applied to other kind of networks, such as genetic network, DNA-protein network, signaling network, metabolic network, food web networks, neuronal networks, species interaction network, etc. The book also provides a chapter dedicated to the discussion of sources of biological network repositories in publicly available databases, which includes a basic introduction to popular and recently used database formats. It can be used as a resourceful reference for the biological network-related researchers.

I am fortunate to be one of the first few readers of the book, and I am excited to recommend it to the researchers who work on biological network analysis. The book is designed to be self-contained, as it includes introductions to the fundamental concepts underlying the graph and graph theory. It can be used as a textbook for advanced graduate courses in bioinformatics. In addition, the book can serve as a resource for graduate students seeking topics in the field of biological network analysis. It can also be used as an excellent reference book for research professionals, who are interested in expanding their knowledge in this topic.

Aidong Zhang
Fellow of ACM and IEEE
Professor of Computer Science
Professor of Biomedical Engineering
University of Virginia, Charlottesville, VA, United States

Preface

The use of graph formalism enables the effective representation of the real-life entities and the complex and mutual interrelationships among the entities. Recently, the use of graphs has become a popular alternative in representing real complex systems, such as social networks, WWW, internet, airline citation, food web, economics, and most importantly, biological networks. Graph modeling of a large class of biological data and systems is growing. The importance of modeling the relationships relates to the consideration that all the biological systems (from the molecular scale to complex organs) are composed of a large set of small components that are strictly interconnected. Therefore, the use of graphs or networks is crucial in current science.

When we started working in the network analysis domain, we strongly realized the need for a good book, covering introductory and advanced topics in biological network analysis. This motivated us to write a handbook that may offer researchers, practitioners, and students a comprehensive guide towards main models, algorithms, and tools available in this area. This book aims to be a vade mecum in this field. While writing this book we tried our best to keep the book suitable for noncomputer science readers too. In it we discuss various graph theoretic and data analytics approaches in practice to analyze networks with respect to tools, technologies, standards, algorithms, and databases available for generating, representing, and analyzing graph data. We referred to recent researches in biological network analysis that might help the budding researchers.

We considered three major biological networks, including gene regulatory networks (GRN), protein-protein interaction networks (PPIN), and human brain connectomes. Keeping in mind the noncomputing readers, we started our discussion with the introductory graph theoretic concepts, which constitute the basis to understand the real-life networks. Real networks are termed as *complex*, due to their unconventional and nontrivial topological properties. We discussed the properties of complex network applicable to real networks. To start empirical study and analysis of various biological networks (that we covered), the sources of various publicly available input networks play an important role. We introduced such sources and the basic details of database and data storage formats. In most chapters, we demonstrated the use of programmes to implement graphs and its analysis using one of the powerful, freely available scripting languages, R.

We believe that this book will be a good starting study material for any research student planning to work in network analysis, and in particular, biological network analysis. To keep the book at a moderate size, we limit discussions on current methodologies and algorithms in use. We encourage interested readers to read original works for more details.

Pietro H. Guzzi
Swarup Roy

Acknowledgment

The authors would like to express deep gratitude to those who were directly and indirectly involved in the writing of this book.

We are grateful to Elsevier, USA for considering the project worthy for publication. Special thanks to Chris Katsaropoulos, Ana Claudia A. Garcia, Kalyanaraman, Prasanna, and their team for translating the dream into reality through the process of proposal finalization, tracking the progress, and final production. It would never be possible to complete the project without their active support and help.

We are thankful to our collaborators: Prof. Jugal Kalita (University of Colorado at CS), Prof. Mario Cannataro, and Prof. Pierangelo Veltri (University of Catanzaro), Prof. Dhruba K. Bhattacharyya (Tezpur University), Dr. Keshab Nath (NIT, Shillong), Dr. Ahed Elmsallati (McKendree University), Dr. Monica Jha (SMIT, Sikkim), Dr. Sazzad Ahmed (Assam Don-Bosco University), Ms. Hazel N. Manners and Mr. Gracious Kharumnuid (NEHU, Shillong), Mr. Partha P. Ray, Sk. Atahar Ali and Ms. Softya Sebastian (Sikkim University), and all the members of the Bioinformatics Lab of the University of Catanzaro for their direct and indirect collaboration and support for writing various chapters. Few of the chapters are the outcomes of various researches conducted with their collaboration during the last few decades.

Few significant experiments on network analysis, reported in this book, were conducted at the network reconstruction and analysis (NetRA) lab, Department of Computer Applications, Sikkim University, India. We are thankful to all the members of the NetRA Lab.

We are thankful to the University of Catanzaro (Italy) and Sikkim University (India) for providing us infrastructure facilities in the University premises for conducting biological network analysis researches. We acknowledge the fund received from the Department of Science and Technology (DST), Govt. of India under DST/ICPS/ Cluster /Data Science/2018 to set-up the NetRA lab and conducting various biological network inference and analysis researches.

While writing the book we referred to a number of related works in this area of research. We hereby acknowledge all the authors, whose research works have been reported in this book.

Last but not the least, we are grateful to the Almighty for everything. We are grateful to our loving family members for their constant support and encouragement. It would never be possible to complete the project successfully without their love and encouragement.

Readers are the most important part of any book. We are grateful to all anonymous readers who consider this book worth reading. Our effort will be successful if it motivates and helps readers in understanding the basics and trends in biological network researches.

Pietro H. Guzzi
Swarup Roy

Abstract

Complex biological systems and their interrelationships are often represented as a graph or network. Therefore, the use of graphs to model biological aspects in computational biology, in bioinformatics, and biomedicine is currently growing, since graphs enable researchers to model relations easily among objects. A main research in this area is represented by the inference and analysis of biological networks from experimental data. Often researchers aim to analyze differences of evolutions among different networks, i.e., network representing different states of the same reality, or networks coming from different species. Consequently a lot of algorithms to analyze and compare networks, have been developed. In particular, the comparison of networks is often performed through network alignment algorithms that rely on graph and subgraph isomorphisms. The book considers three major biological networks, namely, gene regulatory networks (GRN), protein-protein interaction networks (PPIN), and human brain connectomes, and discusses various graph theoretic and data analytics approaches adopted within the last few decades to analyze them with respect to tools, technologies, standards, algorithms, and databases available for generating, representing, analyzing graph data.

Introduction

Contents

"... I can't be as confident about computer science as I can about biology. Biology easily has 500 years of exciting problems to work on. It's at that level."

Donald E. Knuth (1993)

Biology deals with the science of life, whereas computer science models facts into a form that can be dealt with using the help of electronic computing devices. In truth, they are opposite in their actions. Even then, they are brought together to help biologists in handling biological data analysis. Biological experiments are slow and expensive. On the other hand, a computer can handle or process computational problems in less time, and cost-wise, it is less expensive.

In a cell, the outcomes of any cellular activities are not the influence of any particular constituent. Instead, it is a rhythmic and collective role of multiple chemical components inside the cell. It is important to model them mathematically and computationally to derive a sketch on interdependency relationships among the cellular components to elucidate such unknown rhythmic activities within a cell. Biological networks are the graphical and mathematical representation of such interactions or relationships. Graph theory is a mathematical tool that models the pairwise relationships among different objects or entities for the effective analysis. For this obvious reason, graph theory is considered as the best alternative for the formalism of biological network modeling and analysis. Many networks can be considered for describing various biological systems. Protein networks, genetic networks, DNA-protein networks, signaling networks, metabolic networks, food web networks, neuronal networks, species interaction

Biological Network Analysis. https://doi.org/10.1016/B978-0-12-819350-1.00007-4

network, etc., are few biological networks commonly used for the study. We introduce the biological network, how to capture and model them for analysis, next.

1.1 What is a biological network?

Biological systems are composed of little building blocks made up of mutually interacting [3] cellular components. The complex association among various cellular components, such as genes, proteins, metabolites, or even neuronal cells inside brains, is often modeled mathematically as a graph known as **biological network**. The comprehension of such systems is therefore related to the individuation of such elements and their relations. Bioinformatics and systems biology concentrates more on modeling the association among biological molecules on a system-level to improve the knowledge in molecular biology [8]. This approach of knowledge acquisition on a system scale relies on two main pillars: (i) the technological progress leading to the introduction of novel techniques capable of easy, quick, and reliable data generation, and (ii) the development of innovative methodological approaches to apply on the generated data to elucidate novel biological insights. Fig. 1.1 depicts this workflow.

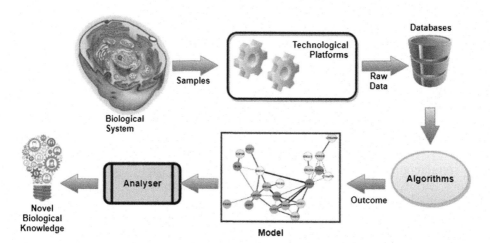

Figure 1.1. Dataflow. The comprehension of biological systems starts with *in vitro* experiments performed in a wet lab. Here, a novel technological platform enables the (high)-throughput production of data related to cell mechanisms. Raw data are then stored in (public) databases. A set of models and algorithms are employed to model and analyze these data.

1.2 Technologies for network data production

The first pillar produces a lot of experimental data to gain insight into the properties of the systems, their properties, and dynamics. For instance, primary PPI data are produced in a wet lab by using different technological platforms. Technologies that enable the determination of protein interactions can be categorized in experiments investigating the presence of physical interactions, and experiments investigating kinetic constants of the reactions. Moreover, based on the number of the interacting partners revealed in a single assay, we can distinguish in technologies that characterize binary relations, such as yeast two-hybrid, and technologies elucidating multiple relations, such as mass spectrometry.

The experiments based on these technologies share a general schema, in which a so-called *bait* protein is used as a test to demonstrate its relations with one or more proteins *preys*. Both single interactions and exhaustive screenings have been realized following this schema. However, an interesting aspect is the reliability of discovered interactions. In particular, each assay can be evaluated on the basis of some ad hoc defined quality measurement.

Considering the human brain connectome of neural cells, the main technologies for data productions are brain imaging techniques, such as magnetic resonance imaging (MRI). Once images have been captured, a set of post-processing techniques are applied to analyze their content and derive brain graphs representing both static and dynamical aspects of the brain.

1.3 Network analysis models

The second pillar has introduced novel tools to build models starting from raw data, and to analyze such models to understand complex systems. Consequently, from a computational science point of view, the need for the introduction of methods and tools for data storage, representation, exchange, and analysis has led the research in such area.

Independent of any network specific application, the flow of data and analysis in this area of research follows standard structure. The process starts with the accumulation of a significant amount of data using high-throughput technologies, such as microarray or next generation sequencing in molecular biology or nuclear magnetic resonance in brain research. Data are then analyzed to build networks, starting from experimental data using network identification methods that result in the building of static or dynamics networks. Networks are mined to elucidate the organization of the biological elements on a system-level scale. Consequently, scientists try to investigate both the global and local organizational principles aiming to discover the difference between subjects or among the healthy and diseased state. After obtaining the networks, the need for the analysis and the comparison of networks of different subjects has led to the development of novel comparison algorithms based on graph and subgraph isomorphism [9].

In case of molecular interactions, after the wet-lab experiments, data are usually collected and preserved in databases [6]. Currently, there exist many publicly available databases that offer the user the possibility to retrieve data easily. Querying interfaces enables both the retrieval of simple information and a particular subnetwork (see Chapter 5 for a more detailed discussion). Many databases can be searched by inserting one or more protein identifiers. The output of such a query is a list of related proteins. Some recent databases offer a semantically more expressive language than simple interaction retrieval, whereas recent research directions are based on the use of a high-level language (e.g., using graph formalism), in suitable graph structures, and search for those by applying appropriate algorithms. Main challenges in this area are (i) expressiveness of the query language that should be able to capture biologically meaningful queries, (ii) efficiency and coverage of the retrieval method, and (iii) simplicity to capture and use results.

Among the others, one of the most used formalism to represent a set of items and their interaction comes from graph theory [4]. Consequently, the use of graphs and networks has become prevalent in many research fields interested in such analysis.

For example, in molecular biology and the so-called omics word, the use of networks for the representation of interactions among proteins is widely used [2], [1]. In interactomic field, proteins, i.e., the interactors, are represented as nodes, whose labels are the identifiers, whereas interactions among proteins are represented as edges linking nodes producing protein-protein interaction networks (PINs). The most straightforward representation uses undirected edges, whereas more refined models use directed and labeled edges to integrate the information about the kind of biochemical association and its direction. Analogously, networks have been largely used to represent the complex mechanism of regulation among genes, yielding to the introduction of gene regulatory networks (GRNs) [7]. In parallel, networks have also been used for the integration of heterogeneous data into a single model [5].

More recently, networks analysis are used in brain research to represent relations among different components of the brain [10]. Contrary to other fields of application, the modeling of the brain, also referred to as connectomics, presents many challenges, since graphs may be used using different scales of views. For instance, nodes may represent neurons and edge their axons, or nodes may be anatomical regions of the brain to understand brain functions and their modifications in case of disease (e.g., for early detection of diseases).

In this book, we consider three types of biological networks: gene expression networks, protein-protein interaction networks and brain connectome networks. We discuss in detail each network models, their properties, the process of generating them, and recent trends and tools in analyzing those networks with an objective to novel biological knowledge, unknown apriori.

1.4 Organization of the book

We organize our book into the following chapters:
- **Chapter 2:** We introduce mathematical graph and properties. A graph is the basis of the entire graph theoretic modeling and analysis of biological networks. We even discuss the R scripting for handling graph data structures, briefly.
- **Chapter 3:** Various algorithms popularly studied in graph theory, such as graph traversal algorithms are discussed. In a biological network, power graph analysis is an important graph analysis method that we discuss with examples. Also, various node centrality measures are introduced and demonstrated with the help of R scripts.
- **Chapter 4:** Real-world networks follow certain special topological properties, which makes them different from the usual graph. Accordingly, they are classified into various network models. We use different models and their properties, and implement them using the R package.
- **Chapter 5:** The sources of three biological network repositories, which are publicly available databases, are discussed. The chapter starts with a basic introduction to popular and recently used database formats. It is a resourceful chapter for the biological network-related researches.
- **Chapter 6:** Gene expression networks have been introduced along with data generation sources for the expression networks. The overall discussion has been divided into two parts, in-silico network inference and post inference analysis. How

gene network modules can be identified and how to rank important genes in an expression network has been discussed in the light of various algorithms. We even discuss various online and offline software tools to carry out gene expression network inference and analysis.

- **Chapter 7:** Protein and their physical interaction networks are vital to establishing true macromolecular connectivity in biological systems. How such interactions can be generated experimentally and predicted computationally has been highlighted. Recently, protein network alignment has gained importance in comparative network analysis for finding evolutionarily conserved proteins, which we include in this chapter. Few of the algorithms dealing with functional protein complex detection is discussed.
- **Chapter 8:** Finally, we introduce brain connectome networks with the input data sources and present trends in brain connectome network analysis.

References

1. Tero Aittokallio, Benno Schwikowski, Graph-based methods for analysing networks in cell biology, Briefings in Bioinformatics 7 (3) (2006) 243–255.
2. Mario Cannataro, Pietro H. Guzzi, Pierangelo Veltri, Protein-to-protein interactions: technologies, databases, and algorithms, ACM Computing Surveys 43 (1) (2010) 1.
3. Nikolay V. Dokholyan, Computational Modeling of Biological Systems. From Molecules to Pathways, Springer Science & Business Media, February 2012.
4. M.C. Golumbic, Algorithmic Graph Theory and Perfect Graphs, Academic Press, New York, 1980.
5. Pietro H. Guzzi, Maria Teresa Di Martino, Pierosandro Tagliaferri, Pierfrancesco Tassone, Mario Cannataro, Analysis of mirna, mrna, and tf interactions through network-based methods, EURASIP Journal on Bioinformatics and Systems Biology 2015 (1) (2015) 1–11.
6. Daniel J. Rigden, Xosé M. Fernández, The 2018 nucleic acids research database issue and the online molecular biology database collection, Nucleic Acids Research 46 (D1) (2018) D1–D7.
7. Swarup Roy, Dhruba K. Bhattacharyya, Jugal K. Kalita, Reconstruction of gene co-expression network from microarray data using local expression patterns, BMC Bioinformatics 15 (7) (2014) S10.
8. Swarup Roy, Pietro Hiram Guzzi, Biological network inference from microarray data, current solutions, and assessments, in: Microarray Data Analysis, Springer, 2015, pp. 155–167.
9. Swarup Roy, H.N. Manners, A. Elmsallati, Jugal K. Kalita, Alignment of protein-protein interaction networks, in: Encyclopedia of Bioinformatics and Computational Biology, Elsevier, 2018, pp. 997–1015.
10. Olaf Sporns, Giulio Tononi, Rolf Kötter, The human connectome: a structural description of the human brain, PLoS Computational Biology 1 (4) (2005) e42.

Preliminaries of graph theory

Contents

Many complex real-life problems, including complex biological systems, are often represented and analyzed using graph theory. The idea of graph is believed to be introduced in the year 1736 by the great Swiss mathematician and physicist, Leonhard Euler, who was asked to find a nice path across the seven Köningsberg bridges; he crossed over each of the seven bridges exactly once. The term "graph" was coined in 1878 by the English mathematician, James Joseph Sylvester [6]. Other than pure mathematics and computer sciences, a plethora of other disciplines, including chemistry, traffic management, electronics, and telecommunications, etc., use it to solve many problems more easily by expressing the problems in terms of graphs. Search giant Google or any GPS service providers use graph theory brilliantly in finding the shortest route from any source station to any destination. Social networks are often represented graphically [4] in order to define social communities of similar interest groups. Regulatory processes inside a cell involve different complex interactions between macromolecular components, such as genes, RNA (Ribonucleic acid), and proteins, commonly referred to as regulatory networks are represented and analysed as graph [5].

To understand biological networks, it is important to have a basic knowledge of graph theory. In this chapter, we explore the basic

Biological Network Analysis. https://doi.org/10.1016/B978-0-12-819350-1.00008-6

concepts and definitions of graphs, types, and representation of graphs, including various operations applicable to graph in general.

2.1 Basic concepts

A graph [3] is a pictorial representation of a set of objects and their association with each other. The objects are popularly termed as nodes or vertices, and the associations are depicted using interconnections between pair of nodes, called edges. Mathematically, graphs are represented as a set of edges and vertices.

Definition 2.1.1 (Graph). A graph $\mathcal{G}$ is a pair of finite set of vertices and edges, $\mathcal{G} = (\mathcal{V}, \mathcal{E})$, such that $\mathcal{V} = \{v_1, v_2, \cdots, v_n\}$ and $\mathcal{E} = \{e_1, e_2, \cdots, e_m\}$. An edge $e_k = (v_i, v_j)$ connects vertices v_i and v_j.

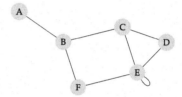

Figure 2.1. An undirected graph with six nodes and eight edges.

In the graph (Fig. 2.1), $\mathcal{V} = \{A, B, C, D, E, F\}$ and $\mathcal{E} = \{(A, B), (B, C), (C, D), (C, E), (E, E), (E, F), (E, D), (F, B)\}$, where edges are an unordered pair of nodes having interconnections among them. Graph $\mathcal{G}$ is termed as **undirected graph**. The node E is connected with itself through loop edge. A graph without my loop structure is called a **simple graph**.

A graph with an ordered pair of nodes, where edges are associated with directions is called a **directed graph** or **digraph**.

Definition 2.1.2 (Directed graph). A directed graph $\mathcal{G} = (\mathcal{V}, \mathcal{E})$ is a set of vertices $\mathcal{V}$ and edges $\mathcal{E}$, such that, for any edge (v_i, v_j) posses direction denoted by arrow. Unlike undirected graph, for any edge $v_i \rightarrow v_j$, the edge $(v_i, v_j) \neq (v_j, v_i)$. The node v_i is called *tail*, and v_j is referred to as *head* of the edge $v_i \rightarrow v_j$. For example, see Fig. 2.2.

Definition 2.1.3 (Path). A path is a sequence of distinct vertices that are connected by edges. In other words, given a set of vertices, $\{v_1, v_2, \cdots, v_k\} \in \mathcal{G}(\mathcal{V})$ is a path if for every pair of vertices v_i and v_{i+1} have an edge $(v_i, v_{i+1}) \in \mathcal{G}(\mathcal{E})$. However, in case of a directed graph, a directed path connects the sequence of vertices

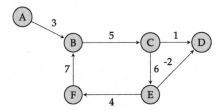

Figure 2.2. Example of digraph showing direction from source node (tail) to destination node (head). The graph is also called **weighted directed graph**, where each edge carries certain weight. Similar weights or costs can be assigned to the edges of any undirected graph.

with the added restriction that all edges are oriented towards the same direction.

In a path, if sequences of vertices are not distinct, it is referred to as a *walk*.

Two nodes, v_i and v_j, are reachable from each other if there is a path that exists between v_i and v_j.

A path is called a *closed path* or *cycle* if two terminal nodes, v_1 and v_k, are connected in a path, i.e., $(v_k, v_1) \in \mathcal{G}(\mathcal{E})$.

In Fig. 2.1, $A - B - C - D$ or $A - B - F - E$ are two different paths, whereas $A \rightarrow B \rightarrow C \rightarrow D$ is a directed path existing in the directed graph (Fig. 2.2), but path $A \rightarrow B \rightarrow F \rightarrow E$ does not exist.

Definition 2.1.4 (Connected graph). A graph is connected when every vertice, v_i, is associated with any edge $e_j \in \mathcal{E}(\mathcal{G})$; there is a path between every pair of vertices. In a connected graph, there are no unreachable vertices. A graph that is not connected is disconnected. See Fig. 2.3.

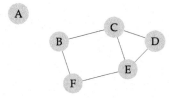

Figure 2.3. A graph becomes disconnected after deleting the edge between A and B.

Definition 2.1.5 (Vertex cut). A set of vertices, $\mathcal{V}' \subseteq \mathcal{G}(\mathcal{V})$, is said to be vertex cut or cut set if $\mathcal{G}(\mathcal{V}) - \mathcal{V}'$ produces a disconnected graph. See Fig. 2.4.

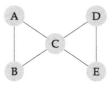

Figure 2.4. The node C is the cut vertex; removing C along with its associated edges leads to the formation of two disconnected graphs.

Definition 2.1.6 (Adjacent node). A node $v_i \in V(\mathcal{G})$ is adjacent to $v_j \in V(\mathcal{G})$ if $(v_i, v_j) \in \mathcal{E}(\mathcal{G})$. Node v_j is said to be *incident* to the edge (v_i, v_j). Two edges are adjacent if they are incident to a common vertex.

The node B, for example, (Fig. 2.4), is adjacent to two nodes A and C. The edge (B, A) and (B, C) are adjacent edges incident to common vertex B.

Definition 2.1.7 (Node neighborhood). Neighborhood, $\mathtt{Neigb}(v_i)$ of a node $v_i \in \mathcal{G}(V)$ is the set of adjacent nodes of v_i:

$$\mathtt{Neigb}(v_i) = \{v_j | (v_i, v_j) \in \mathcal{G}(\mathcal{E})\}. \tag{2.1}$$

Definition 2.1.8 (Node degree). Given a node $v_i \in \mathcal{G}(V)$, the degree (deg) of v_i is the number of edges incident to v_i. In the case of a digraph, the number of edges directed towards v_i is called *indegree*, and total edges going out from v_i towards other nodes is called *outdegree* of v_i.

Theorem 2.1.1. *For a digraph $\mathcal{G} = (V, \mathcal{E})$, the sum of all indegrees is equal to the sum of all outdegrees:*

$$\sum_{i=1}^{|V|} \mathtt{Indeg}(v_i) = \sum_{i=1}^{|V|} \mathtt{Outdeg}(v_i). \tag{2.2}$$

The degree of the node B in Fig. 2.1 is $\deg(B) = 3$. Similarly, the in-degree of the node B in Fig. 2.2 is $\mathtt{Indeg}(B) = 2$ and outdegree is $\mathtt{Outdeg}(B) = 1$. The sum of indegrees is 7, which is the same as the total outdegrees.

In the case of a simple graph, the degree of a node, v_i is nothing but the number of elements in neighborhood of a v_i, i.e., $\deg(v_i) = |\mathtt{Neigb}(v_i)|$.

Theorem 2.1.2 (The handshaking theorem). *Let $\mathcal{G} = (V, \mathcal{E})$ be a graph (with or without loop) with V vertices and $\mathcal{E}$ edges. Then the* *sum of the degrees of node v_i is equal to twice of the number of edges. Consequently, the number of vertices with odd degree is even:*

$$2|\mathcal{E}| = \sum_{i=1}^{|\mathcal{V}|} \deg(v_i). \qquad (2.3)$$

Proof. In a connected graph, any node v_i is adjacent to another node v_j via a common edge (v_i, v_j). Hence, in calculating the degree of v_i and v_j, the edge (v_i, v_j) is considered twice. Thus the sum of degrees of $\mathcal{V}$ vertices is equal to $|\mathcal{E}| + |\mathcal{E}|$.

Since the sum of all degrees is even, it must be that the number of vertices with odd degree is even. $\qquad \square$

Definition 2.1.9 (Regular graph). A graph is said to be a regular graph if every node in the graph possesses the same degree. See Fig. 2.5.

Figure 2.5. Example of a regular graph with four nodes, each with degree three (03).

Definition 2.1.10 (Complete graph). A complete graph, K_n, with n nodes is a regular graph, where every node in the graph is connected to all other nodes directly in the graph. In other words, each node in the graph is the neighbor of all other nodes.

The total number of edges possible in a complete graph is $n(n-1)/2$. The graph shown in Fig. 2.5 is also a complete graph K_4.

Definition 2.1.11 (Subgraph). A graph $\mathcal{G}'$ is a subgraph of $\mathcal{G}$ if $\mathcal{G}'(\mathcal{V}) \subseteq \mathcal{G}(\mathcal{V})$ and $\mathcal{G}'(\mathcal{E}) \subseteq \mathcal{G}(\mathcal{E})$.

Definition 2.1.12 (Induced subgraph). Given a set of vertices, $S \in \mathcal{G}(\mathcal{V})$, the subgraph induced by S, denoted as $\langle S \rangle$, is a subgraph with vertices, S and edges $\mathcal{E}' = \{(v_i, v_j)|(v_i, v_j) \in \mathcal{G}(\mathcal{E}) \wedge v_i, v_j \in S\}$. So, end vertices of any edge in the induced graph must be from S. See Fig. 2.6.

Definition 2.1.13 (Complement graph). A graph $\bar{\mathcal{G}}$ is complement of another graph $\mathcal{G}$ if vertices are same for both the graphs, i.e., $\mathcal{G}(\mathcal{V}) = \bar{\mathcal{G}}(\mathcal{V})$; however, edges are complements such that $\bar{\mathcal{G}}$ contains all the edges not in $\mathcal{G}$, i.e., $\bar{\mathcal{G}}(\mathcal{E}) = \bar{\mathcal{G}}(\mathcal{E})/\mathcal{G}(\mathcal{E})$. See Fig. 2.7.

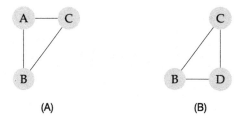

Figure 2.6. Two induced subgraphs derived from the graph given in Fig. 2.5.
(A) Induced graph 1: $\langle S \rangle = \{A, B, C\}$. (B) Induced graph 2: $\langle S \rangle = \{B, C, D\}$.

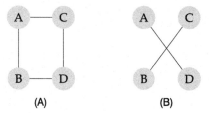

Figure 2.7. Example of complement graph $\mathcal{G}'$ (B) of $\mathcal{G}$ (A).

Definition 2.1.14 (Bipartite graph). A graph is called bipartite if
we can divide the set of vertices into two groups such that every
edge of the graph has one end vertex in one group and another
end in another group. The two groups are called a *partite set*. See
Fig. 2.8.

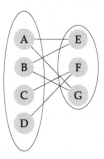

Figure 2.8. Bipartite graph showing two different partitions, where end points of
the edges are separated into two groups.

Definition 2.1.15 (Clique). Given an undirected graph $\mathcal{G}$, a *clique*,
C of $\mathcal{G}$ is an induced subgraph of $\mathcal{G}$ induced by C, which is complete. In other words, C a subset of the vertices $C \subset \mathcal{G}(\mathcal{V})$ such that
every two distinct vertices in C are adjacent. See Fig. 2.9.

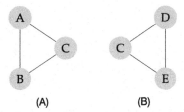

Figure 2.9. Two cliques in the graph in Fig. 2.4. (A) Induced graph 1: $\langle S \rangle = \{A, B, C\}$. (B) Induced graph 2: $\langle S \rangle = \{C, D, E\}$.

2.2 Data structure for representing graphs

Graphs can be represented in the computer memory using *array* and *list* data structures.

2.2.1 Array representation

The sequential representation of a graph using an array data structure uses a two-dimensional array or matrix called *adjacency matrix*.

Definition 2.2.1 (Adjacency matrix). Given a graph $\mathcal{G} = (\mathcal{V}, \mathcal{E})$, an adjacency matrix, say Adj is a square matrix of size $|\mathcal{V}| \times |\mathcal{V}|$. Each cell of Adj indicates an edge between any two vertices or nodes:

$$Adj[i][j] = \begin{cases} \omega, & \text{if } (v_i, v_j) \in \mathcal{G}(\mathcal{E}) \\ 0, & \text{otherwise,} \end{cases} \qquad (2.4)$$

where ω is the weight of the edge between the nodes v_i and v_j. In the case of an unweighted graph, ω is considered as 1, whereas for weighted graph it may be any value according to the problem in hand. See Fig. 2.10.

Adjacency matrices of undirected graphs are symmetric, where $Adj[i][j] = Adj[j][i]$, for i, j. In other words, we may say that Adj and its transpose Adj' is the same. Unlike undirected graph, digraph produces asymmetric matrix.

Finding degree of a node

One of the important operations on a graph is finding the degree of a given node. From the adjacency matrix, it is easy to determine the connection of any nodes. The degree of a node in an undirected graph can be calculated as follows:

$$\deg(v_i) = \sum_{j=1}^{n} Adj[i][j], \qquad (2.5)$$

	A	B	C	D	E	F
A	0	3	0	0	0	0
B	0	0	5	0	0	0
C	0	0	0	1	6	0
D	0	0	0	0	0	0
E	0	0	0	-2	0	4
F	0	7	0	0	0	0

(A)

	A	B	C	D	E	F
A	0	1	0	0	0	0
B	1	0	1	0	0	1
C	0	1	0	1	1	0
D	0	0	1	0	1	0
E	0	0	1	1	1	1
F	0	1	0	0	1	0

(B)

Figure 2.10. Adjacency matrix representation of weighted digraph and undirected graph. (A) Adjacency matrix of the weighted diagraph in Fig. 2.2. (B) Matrix representation of undirected graph in Fig. 2.1.

where values in the ith row in the adjacency matrix indicates the connections to n different nodes from the node i in the graph. Similarly, in the case of digraph, the indegree and outdegree of a node can be calculated as follows:

$$\text{indeg}(v_i) = \sum_{j=1}^{n} Adj[j][i] \text{ and } \text{outdeg}(v_i) = \sum_{j=1}^{n} Adj[i][j]. \quad (2.6)$$

2.2.2 List representation

Array data structures are easy to access and fast in traversing. However, for a large graph, it is not always feasible to use adjacency matrix representation, due to large memory requirements. It is even more ineffective if a graph contains more nodes with relatively few connections or edges (sparse graph); this leads to the formation of a *sparse matrix*. To overcome such situation, list representation is an effective alternative for memory representation of large and dense graphs. An advantage of list representation is that it can be used for dynamic graphs, where vertices and edges are growing and shrinking with time. It is commonly implemented in any programming languages as an array of a singly-linked list. The size of the array is the number of vertices in the graph. Each singly linked list keeps track of the neighbors of a vertex. In the case of a weighted graph, the weights of an edge between a pair of vertices are stored in the nodes of singly-linked list itself as a separate entry together with vertex level. It is easy to calculate the degree of a vertex by looking into the number of nodes in the list of the vertex. For example, the degree of the vertex **C**, which is four (04), can easily be calculated by finding the length of the list headed by C, as given in Fig. 2.11.

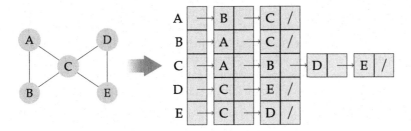

Figure 2.11. Linked list representation of an undirected graph.

2.3 Trees

A tree is a popular graph data structure used in computer science for hierarchical representation of data. It is a minimally connected graph with no cycle, also termed as an *acyclic graph.* The acyclic means that there is no alternative path or edge from any nodes to reach any other nodes in the graph. It can have an only one-to-many relationship down the hierarchy, and not the reverse. A tree may be **rooted** and **unrooted**.

2.3.1 A rooted tree

In a rooted tree, there is a specific root from where the tree grows. Unlike a real tree, a tree in graph theory is usually represented as an inverted tree starting with a root. Each node is attached to one or more child nodes. However, it can only have a single parent. A typical general rooted tree structure is shown in Fig. 2.12. A tree is organized as levels. A root is always at level 0, and subsequent children in the hierarchy are arranged in different higher levels according to their relationships. As shown in Fig. 2.12, Root **A** is in level 0, and **B** and **E** is its immediate children or offsprings. Child **C**, **D**, **F**, and **G** in level 2 are called leaf nodes, as there are no children linked with the nodes. Following the real-life family hierarchy, node **B** and **E** are called siblings of each other. On the other hand, node **C** and **F** are not siblings, as they are derived from two different parents.

Mathematically, a rooted tree can be represented as a finite set of nodes such that the followings hold:
- There is a specially designated node called the *root.*
- The remaining nodes are partitioned into $n \geq 0$ disjoint sets $\mathcal{T} = \{T_1, T_2, \cdots T_n\}$, where each $T_i \in \mathcal{T}$ is a tree itself. $T_1, T_2, \cdots T_n$ are called the subtrees of the root.

Depending on the organization of the nodes of a tree, it may be ordered or unordered. An ordered tree is a rooted tree, in which the order of the children at every node is specified. It is an organized tree in terms of how nodes are organized. A tree is unordered if the ordering of the nodes are not important and has an arbitrary order. We may consider the tree given below as an ordered tree, where every node is a lexicographically prior to its children.

Some of the basic terminology related to a rooted tree is listed below.

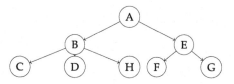

Figure 2.12. A general rooted tree with root and child nodes. There are three levels (0 to 2) in the tree.

Definition 2.3.1 (Null tree). A tree is called a null or empty tree if no nodes are present in the tree.

Definition 2.3.2 (Degree of a tree node). The degree of a node is the number of subtrees of the node. The degree of **A** is 2, where the degree of **D** is 0 (Fig. 2.12).

Definition 2.3.3 (Leaves). Nodes with no children are called leaves. In another words, a node with degree 0 is the leaf node.

Definition 2.3.4 (Path). A path in a tree is a list of distinct nodes, in which successive vertices are connected by edges in the tree. For example, $\{A, B, D\}$ in the above tree is a path. The defining property of a tree is that there is precisely one path connecting any two nodes.

Definition 2.3.5 (Tree height). Depth or height of a tree is the number of levels present in the tree. If a tree has root in level zero, then the height of the tree is the $maximum level + 1$. In our example tree, the depth of the tree is $2 + 1 = 3$.

2.3.2 Binary tree

A binary tree is a rooted tree, where each node has at most two children. In other words, each parent can have 0 to 2 children only. Formally, it can be defined as a finite set of nodes, which is either *empty*, or consists of a root and two disjoint binary trees called the *left subtree* and *right subtree*. For example, a tree in Fig. 2.14A is a binary tree.

Though the binary tree is a kind of tree, based on its definition, distinction exists between them. There is no tree having zero nodes, but there is an empty binary tree. Secondly, the two trees given in Fig. 2.13 are not the same binary trees. The first one has an empty right subtree, whereas the second has an empty left subtree. If the above are regarded as trees, then they are the same despite the fact that they are drawn slightly differently.

Figure 2.13. Two different binary trees.

Kinds of binary trees

Based on the constraints imposed on any binary tree, it can be divided into following specialized binary trees:

Definition 2.3.6 (Full binary tree). A binary tree is said to be a full or strictly binary tree if each node contains either two children or no children; the permissible degree of any node is strictly either 0 or 2.

Definition 2.3.7 (Complete binary tree). A complete binary tree is where every level must be filled with maximum possible nodes in that level. The last level may be partially filled as far left as possible. In other words, in a binary tree with depth k is a complete binary tree if the lowest $k - 1$ levels of the tree are filled, and level k is partially filled from left to right.

Definition 2.3.8 (Perfect binary tree). In a perfect binary tree, all the leaves are at the same level. A perfect binary tree is a full tree of depth k and having $2^{k+1} - 1$ nodes.

Some properties of binary tree

Few interesting properties of a binary tree are listed below.

Theorem 2.3.1. *The maximum number of nodes on level L of a binary tree is* 2^L, $L >= 0$.

Proof. This can be proved by the method of induction. For root, $L = 0$, the number of nodes is $2^0 = 1$. Assume that the maximum number of nodes in any level L is 2^L. Since in binary tree every node can have maximum 2 children, therefore the next level

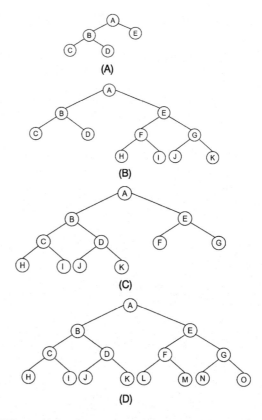

Figure 2.14. Different kinds of binary trees. (A) A strictly binary tree. (B) A fully binary tree and not a complete tree. (C) A complete binary tree with depth, $k = 4$. (D) A perfect binary tree and a full binary tree.

$(L + 1)$ would have twice nodes of level L, i.e., 2×2^L, which is nothing but 2^{L+1}.   □

Theorem 2.3.2. *The maximum of number of nodes in a binary of height k is $2^{k+1} - 1$.*

Proof. A tree with height k contains $k - 1$ level (root at level 0). By Theorem 2.3.1, the maximum number of nodes at level L is 2^L. Therefore, maximum number of nodes in a tree of height k is the sum of the geometric series.

$$N = \sum_{L=0}^{k-1} 2^L = \frac{2^{k+1} - 1}{2 - 1} = 2^{k+1} - 1$$

□

Theorem 2.3.3. *If N be the total number of nodes, then the height of the tree is at most N − 1 and at least* $\lceil \log_2 N \rceil$.

Proof. In case of extreme skewed tree each level contains one node. Therefore, the height of the tree become $N - 1$ (root at 0 level).

However, in case of perfect binary tree of height k, the maximum number of nodes (Theorem 2.3.2) is

$$N = 2^{k+1} - 1$$
$$k + 1 = \log_2 N + 1$$
$$\text{therefore, } k = \lceil \log_2 N \rceil$$

$\square$

2.3.3 An unrooted tree

A tree with no specific root is referred as an unrooted tree. It is a connected undirected acyclic graph. An unrooted tree is most commonly used in bioinformatics to represent evolutionary history of living organisms; an example of such a tree is the *phylogenetic tree* [7]. An unrooted phylogenic tree considers no knowledge of a common ancestor, it indicates only location of the taxa (group of one or more populations of an organism) to represent its relative relationships among the taxa. Usually, an unrooted tree is converted to a rooted tree by assuming a dummy root before processing the tree. See Fig. 2.15.

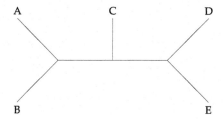

Figure 2.15. Example of unrooted tree with no specific root.

2.3.4 Representation of a tree

Like a graph, a tree can also be represented using arrays and linked-list data structures. List representation is more common for a general tree. However, due to a nice property of parent-child ordering in a binary tree, it can efficiently be represented using arrays. In addition, a simplistic parenthesis representation of a tree is also possible.

Newick format

Any tree can easily be represented in linear form as nested parenthesis [1]. Every root of a subtree will be the first element followed by the left sub-tree and right-subtree covered within the opening and closing parenthesis. Every subtree is recursively expanded using the same parenthesization downwards the hierarchy of the tree. For example, we may represent the rooted tree in Fig. 2.12 as follows:

$$(A, (B, (C, D)), H), (E, F, G)));$$

In the case of an unrooted tree, due to the absence of specific root, it can be represented in many alternative forms. In the case of the unrooted tree, the convention is simply to consider the root of the tree arbitrarily and report the resulting rooted tree. For example, the unrooted tree in Fig. 2.15 can be represented as either of the many possible forms. Few of them are listed below:

$$(((A, B), C, (D, E)));$$
$$((A, (B, C)), (D, E));$$
$$((A, B), (C, D), E);$$

List representation

In linked list representation a node must have a varying number of fields, depending upon the number of branches. However, it is easier to handle a tree, where the node size is fixed. Every tree node contains *info* part to store data or useful information. One or more *link* parts is to store the pointers to its child nodes. In the case of a binary tree, the number of links is fixed, i.e., two. Another alternative linked list-based representation is **left child-right sibling** representation to store in each node the data, a pointer to left child and a pointer to the list of right siblings (Fig. 2.16B).

Array representation

A binary tree with N nodes can be represented sequentially using array. Given any node with index i, $1 <= i <= N$, we can represent the child parent relationship as follows with regards to array index:

- the parent of i is at $i/2$ position if $i \neq 1$. If $i = 1$, i is at the root and has no parent;
- the left child of i is at $2i$ if $2i <= N$. If $2i > N$, then i has no left child;
- the right child of i is at $2i + 1$ if $2i + 1 <= N$. If $2i + 1 > N$, then i has no right child.

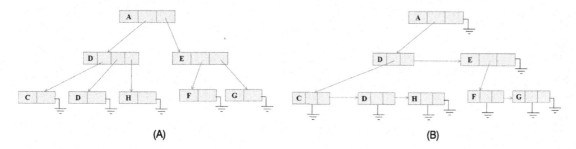

(A)   (B)

Figure 2.16. Linked list representation of the tree in Fig. 2.12. (A) List representation. (B) Left child-right sibling representation.

Accordingly, the array representation of the binary tree in Fig. 2.14A can be represented as follows (Fig. 2.17). It is faster and easy to access in comparison to list representation.

1	2	3	4	5	6	7
A	B	E	C	D	*	*

Figure 2.17. Array representation of a binary tree. Size of the array should be $2^k - 1$ for k depth tree. * indicates null value.

2.4 Implementing graphs in R

Due to ease of implementation, the R language is used popularly for any statistical and mathematical computations. R is a free, open-source software under the terms of the Free Software Foundation's GNU General Public License. R supports multiple operating system platforms, including Windows, Linux, and MacOS. To work with graph or network, *igraph*[1] is a powerful library, which is freely downloadable and implemented in R, Python, and C programming environments. To generate a graph in R using igraph [2], one needs to install the package first. Any graphs can be generated using various igraph library functions. However, a graph can also be generated using the user input file. Below we present snapshots of series of R codes to demonstrate the use of igraph in R. We first create an input file Adj.csv (Table 2.1) corresponding to the graph given in Fig. 2.11 by storing two column edge data, and input it to R functions. The R script for creating graphs and the corresponding outputs are shown in Fig. 2.18 and Fig. 2.19 respectively.

[1]http://igraph.org.

Table 2.1 Adj.csv: adjacency matrix for demonstration.

A	B
A	C
B	C
C	D
C	E
D	E

```
> install.packages("igraph")//installation igraph package
> library(igraph) //including the igraph library

//Reading input file Adj.csv in to the handler link
> links <- read.csv("ADJ.csv", header=F, as.is=T)

//Creating an undirected graph or network from input data
> net <- graph.data.frame(links, directed=F)
> plot(net)          //Drawing the undirected graph

//Creating an directed graph or network from input data
> net <- graph.data.frame(links, directed=T)
> plot(net)      //Drawing the directed graph
```

Figure 2.18. Sample R script to show the use of igraph library for network analysis.

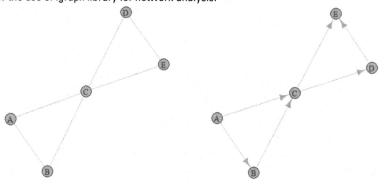

Figure 2.19. Outcome of the R code in Fig. 2.18.

2.4.1 More graphs in R

Different types of graph can be generated using R. Few are demonstrated in Table 2.2.

Table 2.2 Generating different graphs using R.

R Code	Output
Empty Graph	

```
net=graph.empty(n=30, directed=TRUE)
plot(net, vertex.size=10, vertex.label=NA)
```

Complete Graph

```
net=graph.full(10)
plot(net, vertex.size=10, vertex.label=NA)
```

Directed Graph with loop

```
net=graph(edges=c(1,2,2,3,3,1,3,3,3,2),
                   n=3, directed=T)
plot(net)
```

Tree

```
tree2 <- make_tree(10, 3)
plot(tree2, layout=layout_as_tree)
plot(tree2, layout=layout_as_tree(tree2,
        root=c(1,11),rootlevel=c(2,1)))
```

2.5 Summary

The graph is a useful tool for the mathematical representation of any complex relationship among different entities. Biological relationships are commonly represented using graphs for better understanding, and ease of computer implementation. Once represented as a graph, the biological networks can be analyzed using different standard graph analysis algorithms or customized as per the problem in hand. In the next chapter, we discuss some of the algorithms commonly used for graph analysis.

References

1. Gabriel Cardona, Francesc Rosselló, Gabriel Valiente, Extended Newick: it is time for a standard representation of phylogenetic networks, BMC Bioinformatics 9 (1) (2008) 532.
2. Gabor Csardi, Tamas Nepusz, et al., The igraph software package for complex network research, InterJournal, Complex Systems 1695 (5) (2006) 1–9.
3. Narsingh Deo, Graph Theory With Applications to Engineering and Computer Science, Courier Dover Publications, 2017.
4. Keshab Nath, Swarup Roy, Detecting intrinsic communities in evolving networks, Social Network Analysis and Mining 9 (1) (2019) 13.
5. Georgios A. Pavlopoulos, Maria Secrier, Charalampos N. Moschopoulos, Theodoros G. Soldatos, Sophia Kossida, Jan Aerts, Reinhard Schneider, Pantelis G. Bagos, Using graph theory to analyze biological networks, BioData Mining 4 (1) (2011) 10.
6. James Joseph Sylvester, On an application of the new atomic theory to the graphical representation of the invariants and covariants of binary quantics, with three appendices, American Journal of Mathematics 1 (1) (1878) 64–104.
7. Jin Xiong, Essential Bioinformatics, Cambridge University Press, 2006.

3

Graph analysis

Contents

Graph analysis refers to the set of algorithms for analyzing properties of a graph. The main algorithms for graph analysis and traversal are introduced in this chapter. We start by discussing graph traversal, i.e., the problem of visiting the nodes of a graph. Many algorithms of graph analysis are based on an extension or modification of graph traversal algorithms. We then discuss the ways to find a shortest path among nodes in a graph. We also consider the compression of graphs through power graph analysis and the discovery of communities (i.e., a subgraph having some given properties). Node centrality measures, used for finding vital nodes are also discussed. We demonstrate the use of various graph traversal and analysis methods using R libraries.

3.1 Traversing a graph

Despite the existence of many algorithms for graph analysis, it should be noted that a large class of them employs graph traversal algorithms. Graph traversal, also known as graph search, refers to the process of enumerating (or visiting, or checking, or modifying), all the nodes of the graph. There exist two ways of performing traversals, which differ by the order in which vertices are visited.

Graph traversal algorithms start by visiting a node of the graph, termed as source or initial node (**initial or source node**). Next, it explores all the nodes of the graph (assuming that the graph is connected) and returning the list of visited nodes. Once a node has been traversed, that node is marked or coloured as visited (to guarantee the termination of the algorithm). Then the algorithm selects one of its adjacent ones, and the algorithm ends when all the nodes have been visited. The selection of the adjacent one will define the properties of the traversal, and based on this selection, two different techniques have been developed, namely **breadth first search** and **depth first search**, as depicted in Fig. 3.1.

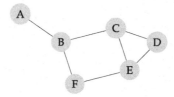

Figure 3.1. Starting from Node A, a DFS traversal will visit the graph in this order: A, B, C, D, E, F; a BFS traversal will visit the graph in this order: A, B, F, C, D, E.

Graph traversal techniques have been developed both for ordered and unordered graphs. In the following discussions, we assume that the graphs are unordered. If the graph has many connected components, then the traversal will return to the list of the nodes of the connected component to which the start node belong. Consequently, a generic algorithm for graph traversal is represented in Fig. 3.2. The algorithm is based on three main concepts: a collection of visited nodes, a collection of unvisited nodes and the choice of a node from this list. The algorithm initially marks all nodes unvisited. Then it starts from the starting node. After that starting node has been visited, it considers its adjacent node, and iteratively visits all the nodes. The implementation of the collection `toVisitNodes` determines the characteristics of the graph traversal.

```
procedure Visit (Graph H, Node startnode,
  Collection toVisitNodes, Collection visitednodes )
Mark all nodes as not visited;
put (startNodes,toVisitNodes)

while(toVisitNodes is not empty ) do
   {
      x = pop(toVisitNodes);
      if ( x has not been visited )
      {
         mark x as visited;            // Visit node x !
         { for each node y adjacent to x
           if ( y has not been visited )
             push (y, toVisitNodes); }
```

Figure 3.2. A general algorithm for realizing graph traversal.

3.1.1 Breadth first search (BFS)

A possible choice may be the *first-in-first-out* by using a **queue** while implementing BFS. In queue based implementation after visiting a node, all the adjacent ones are added to the queue. Then a first adjacent is visited, and its adjacent are added to the queue. Before visiting these nodes, all the adjacent nodes of the first node is visited. It is obvious that all the nodes that are a distance k from it are visited before any node at a distance $k + 1$. This traversal is called breadth-first search and is depicted in Fig. 3.3.

The time complexity can be expressed as $\mathcal{O}(|\mathcal{V}| + |\mathcal{E}|)$, where $|\mathcal{V}|$ is the number of the nodes, and $|\mathcal{E}|$ is the number of the edges. In the worst-case, every vertex and every edge will be explored. The formula clearly shows that the time complexity may vary depending on the sparsity of the input graph (see [10] for a deep discussion of the complexity in many different cases). This algorithm requires the building of the queue, then it requires the space for storing the graph, and a supplementary space $\mathcal{O}(|\mathcal{V}|)$ for the queue.

The possibility of speed-up of BFS through parallel algorithms has been explored in the past [24]. BFS is used in many other algorithms such as the following:

- Finding the shortest path between two nodes (see next sections);

```
procedure Breadth First Search (Graph H, Node startnode,
  Queue toVisitNodes, Collection visitednodes )
Mark all nodes as not visited;
put (startNodes,toVisitNodes)

  while(toVisitNodes is not empty ) do
    {
        x = dequeue(toVisitNodes);
        if ( x has not been visited )
        {
            mark x as visited;          // Visit node x !
          { for each node y adjacent to x
            if ( y has not been visited )
              enqueue (y, toVisitNodes); }
```

Figure 3.3. Breadth-first search algorithm using queue.

- Computing the maximum flow in a flow network (Ford–Fulkerson algorithm [16]);
- Testing bipartiteness;
- Cheney's algorithm [8] for managing garbage collection in shared memory.

3.1.2 Depth-first search (DFS)

An alternative way to manage the list of nodes to be visited is the *last-in first-out*. In this way, when a node is visited, all the adjacent nodes are added into the list. Then one of them is visited, and its adjacent nodes are inserted. At this point, the algorithm selects one of them and this process iterate until there exists at least one more adjacent node. It is evident that the algorithm for each step $i + 1$ visits a node at distance $i + 1$, and stores the node at distance i. This traversal, called depth-first search, uses a **stack**, as depicted in the algorithm (Fig. 3.4).

Similarly to BFS, the time complexity for DFS is $\mathcal{O}(|\mathcal{V}| + |\mathcal{E}|)$, where $|\mathcal{V}|$ and $|\mathcal{E}|$ are, respectively, the number of nodes and edges. The space needed for the execution is related to the storing of the stack; therefore it needs additional $\mathcal{O}(|\mathcal{V}|)$ space as well as the set of already-visited vertices.

Few of the well-known applications of DFS are:

- Finding connected and strongly connected components;
- Finding the bridges of a graph;
- Finding biconnectivity in graphs.

```
procedure Depth First Search (Graph H, Node startnode,
  Stack toVisitNodes, Collection visitednodes )
Mark all nodes as not visited;
push (startNodes,toVisitNodes)
while(toVisitNodes is not empty ) do
    {
        x = pop(toVisitNodes);
        if ( x has not been visited )
        {
            mark x as visited;           // Visit node x !
          { for each node y adjacent to x
            if ( y has not been visited )
              push (y, toVisitNodes); }
```

Figure 3.4. Depth-first search algorithm using stack.

3.2 Graph traversal at a glance

The followings hold in connection to graph traversal at a glance:

- BFS considers all of the adjacents of a node before visiting other nodes. DFS visits all the children nodes (i.e., neighbors of the adjacent node) before visiting neighbors.
- BFS and DFS are based on a similar algorithm, and they differ in that BFS uses a queue data structure, whereas DFS uses a stack.
- BFS needs of a larger amount of memory in general, because it expands all children of a vertex and stores them in memory. DFS has much lower memory requirements than BFS, because it iteratively expands only one child until it cannot proceed anymore and backtracks after that.
- Both have the same time complexity.
- BFS is used to find the shortest path in a graph, whereas DFS is used to find a path among nodes (i.e., determining if every pair of vertices can be connected by two disjoint paths).
- The choice between both algorithms depends on the context.

3.3 Shortest paths in a graph

One of the most interesting problems in a graph is finding the *best* path connecting two nodes. *Best* means the shortest (in terms of geodesic distance) for unweighted graphs, or a path such that the sum of the weights of its constituent edges is minimized for weighted graphs. The shortest path problem may be defined for both directed and undirected graphs. Here, we discuss paths in undirected graphs, and the reader may easily extend the discussion for directed ones [9].

In introducing the problem, we briefly recall from Chapter 2 that a path in a graph $\mathcal{G}$ is an ordered sequence of adjacent nodes $P = \{v_1, v_2, \ldots, v_n\}$ such that there always exists an edge connecting v_j, v_{j+1} for all $j \in 1, n-1$. The path P connects nodes v_1 and v_n. The number of nodes in P is called the length of the path P. The ordering of nodes in P determines the structure of the path. Given any pair of nodes of a graph, if there exists a path between them, then they are connected, and in general, two nodes may be connected by multiple paths of different length. Given an unweighted graph $\mathcal{G}$ and two nodes, v and w, the *shortest path* $\mathcal{SP}$ is the path connecting v and w with the minimum length. Similarly, for weighted graphs, given a real-valued weight function $f : \mathcal{E} \rightarrow \mathbb{R}$, the shortest path from v_1 to v_n is the path $P = (v_1, v_2, \ldots, v_n)$ that minimizes the sum $\sum_{i=1}^{n-1} f(e_{i,i+1})$ over all possible paths. When each edge in the graph has unit weight or $f : \mathcal{E} \rightarrow \{1\}$, this is equivalent to finding the path with fewest edges.

The shortest path problem has three main variants:

- The single-source shortest path problem, in which the algorithms aim to find all the shortest paths from a given source node v to all other vertices in the graph.
- The single-destination shortest path problem, in which algorithms aim to find shortest paths from all nodes in the directed graph to a single destination node v.
- The all-pairs shortest path problem, in which we have to find the shortest paths between every pair of nodes in the graph.

We point out that each one of these variants have been resolved by efficient algorithms than the naive approach of running a single-pair shortest path algorithm on all relevant pairs of vertices [10].

3.3.1 Dijkstra's shortest path first algorithm

Dijkstra's shortest path first algorithm (or Dijkstra's algorithm) finds the shortest paths between nodes in a graph, and it has been used both to find the shortest path between two given nodes, and to find shortest paths from a given source to all other nodes [12]. The simplest version of the algorithm (see Fig. 3.5) assumes that each edge has a nonnegative weight. The behavior of the algorithm may be described as follows.

Initially, two node-sets are created: Q, the list of unvisited nodes, and *dist* the array of distances from the initial node u. For each node v, let the distance from u be ∞. Then the algorithm adds

```
procedure Dijkstra(Graph G, Node source):
 create node set Q
     for each vertex v in G:
         dist[v] = Infinite
         prev[v] = None
         add (v,Q)
     dist[source] = 0

     while Q is not empty:
         let i argmin dist[u] in Q
         u = Q[i]
         remove u from Q
         for each neighbor w of u in Q:
                         newdist = dist[u] + length(u, v)
                 if (newdist < dist[v]):
                                 dist[v] = newdist
                                 prev[v] = u
         return dist[], prev[]
```

Figure 3.5. Pseudocode of the Dijkstra's algorithm.

all the nodes to (Q) and set the initial node as current. Starting from the current node, it considers all of the unvisited neighbors in Q. For each of them, it calculates their tentative distances from the initial node through the current node. If the newly calculated distance is lower than the previous one, it will assign the smaller one; otherwise, it does not change the current value. Then (when planning a route between two specific nodes) if the shortest tentative distance among the two nodes is infinity, or if the destination node has been reached (i.e., marked as visited), the algorithm ends. If the destination node has been marked visited (when planning a route between two specific nodes), or if the smallest tentative distance among the nodes in the unvisited set is infinity (when planning a complete traversal, which occurs when there is no connection between the initial node and remaining unvisited nodes), then stop. The algorithm has finished. Otherwise, it selects the unvisited node that is marked with the smallest tentative distance, set it as the new "current node", and go back.

The complexity of the Dijkstra's algorithm is related to the implementation of data structures needed for the processing of nodes, such as the node-set Q. The simplest upper bound of running time may be expressed as the time needed to process all the nodes $|\mathcal{V}|$ and edges $|\mathcal{E}|$, giving $\mathcal{O}(|\mathcal{V}| + |\mathcal{E}|)$. A more detailed analysis of the running time is related to the implementation of the node-set Q. The algorithm performs two main operations in Q, the extraction of the node with the minimum distance from the current, and the remotion of a node from Q. Let T_{min} be the time for performing the first operation), and let T_{del} be the time for the second one. Then the time complexity may be expressed as $\mathcal{O}(|\mathcal{V}| * T_{del} + |\mathcal{E}| * T_{min})$. In the case of the use of a list, this gives that the time complexity as $\mathcal{O}(|\mathcal{V}| * |\mathcal{V}| + |\mathcal{E}|) = \mathcal{O}(\mathcal{V}^2)$, since the extraction of the minimum needs the exploration of the whole list, whereas the elimination has a constant time.

3.3.2 Handling negative edge weights: the Bellman–Ford algorithm

The Dijkstra's algorithm is relatively fast and easy to implement, but it lacks the processing of graphs with negative edge weights and possibly negative cycles. To address these limitations, Richard Bellman [3] and Lester Ford Jr [17] published the same algorithm for finding shortest paths in graphs with negative edge weights, sometimes called the Bellman–Ford algorithm.

The Bellmann–Ford algorithm iteratively calculates minimum distances, and for each step it updates the distances until they eventually reach the solution. All the distances are initially set to ∞. Similarly to Dijkstra's algorithm, for each step, these distances are calculated and eventually updated. The main difference is that the Bellman–Ford algorithm considers all the edges for each level, whereas Dijkstra's algorithm uses a queue to greedily select the closest node that has not yet been processed, and it considers the outgoing edges.

The pseudocode for the algorithm is depicted in Fig. 3.6. The computational complexity of the algorithm is $\mathcal{O}(|\mathcal{V}| * |\mathcal{E}|)$.

3.3.3 A dynamic programming approach: the Floyd–Warshall algorithm

The term *dynamic programming* refers to a set of methods for solving problems using both mathematical optimization and recursive computer programming. Dynamic programming was introduced by Richard Bellman in the 1950s. The core idea of dynamic programming, as discussed in the paper by Bellman [4], is to simplify a complex problem into simpler ones recursively until they may be solved singularly. Then the whole solution is derived by merging all the sub-solutions. When a problem may be recursively decomposed into smaller ones, and the solution may be obtained by solving all the subproblems, then the problem has an optimal substructure. It should be noted that some problems cannot be decomposed in an optimal way; therefore dynamic programming does not apply to all the problems.

```
procedure BellmanFord( Graph G, node source)

    for each node v in G:
        dist[v] = infinite
        prec[v]   null
    // prec is used to store the path, dist is used to store distances

    dist[source] := 0

    // Step 2: Processing edges

    for each node v:
        for each edge (u, v) in G with weight w:
            if dist[u] + w < dist[v]:
                dist[v] = dist[u] + w
                prec[v] := u

    // Analysis of negative-weight cycles
    for each edge (u, v)in G with weight w:
        if dist[u] + w < dist[v]:
            error "Graph contains a negative-weight cycle"
```

Figure 3.6. Pseudocode for the Bellman–Ford algorithm.

The Floyd–Warshall algorithm [15] finds the shortest paths in a weighted graph with positive or negative edge weights (but with no negative cycles) using dynamic programming. The original version of the algorithm finds only the length of paths between all pairs of vertices, whereas the paths have been added with some modifications to the algorithm.

Let us consider a graph $\mathcal{G} = (\mathcal{V}, \mathcal{E})$, where $|\mathcal{V}| = N$, a numbering function that assigns to each node a value in $\{1, 2, \ldots, n\}$ and a function shortestPath(i, j, k) that returns the shortest possible path from node $i \in \mathcal{V}$ to $j \in \mathcal{V}$ using vertices only from the set $1, \ldots, k$ for building the path between i and j. The goal of the algorithm is to find the shortest path for all the possible pairs (i, j) considering any node in $\mathcal{V}$, i.e., $k = n$.

For each of these pair of vertices, the $shortestPath(i, j, k)$ may use or not k as explained in the following:

- a path that does not use k, i.e., it uses vertices in the set $\{1, \ldots, k - 1\}$, that may be defined as $shortestPath(i, j, k - 1)$,
- a path that uses k, i.e., a path that may be splitted into a path from i to k, and then from k to j—both only using intermediate vertices in $\{1, \ldots, k - 1\}$—may be defined as $shortestPath(i, k, k - 1)$, and $shortestPath(k, j, k - 1)$.

Consequently, the best path from i to j would be the concatenation of the shortest path from i to k, (using intermediate vertices in $\{1, \ldots, k-1\}$), and the shortest path from k to j (using intermediate vertices in $\{1, \ldots, k-1\}$). The previous definition may be recursively formulated as follows:

$$shortestPath(i, j, k) = min \begin{cases} shortestPath(i, j, k-1) \\ shortestPath(i, k, k-1) \\ +shortestPath(k, j, k-1), \end{cases} \quad (3.1)$$

and the base case of the recursion is $shortestPath(i, j, 0) = w(i, j)$, i.e., the shortest path between two adjacent nodes is the weight of the connecting edge.

Once we define the formula that is the core of the Floyd–Warshall algorithm, we may explain the main steps.

The algorithm computes the shortest path among all the vertices in an incremental way, i.e., for each pair (i, j) it computes all the $shortestPath(i, j, k)$ varying k from 1 to N, and the implementation is presented in the fragment of code given in Fig. 3.7.

```
procedure Floyd-Warshall Algorithm(Graph G):

dist[|V|][|V|] is the array of distances
for each (i,j)
    dist[i][j]=infinity
//Base Case of the Recursion
for each edge (i,j)
    dist[i][j] = w(i,j)

for each node i
    dist[i][i] = 0

for k from 1 to |V|
    for i from 1 to |V|
        for j from 1 to |V|
            if (dist[i][j] < dist[i][k] + dist[k][j])
                dist[i][j] = dist[i][k] + dist[k][j]
            end if
```

Figure 3.7. Implementation of the Floyd–Warshall algorithm.

The computational complexity of the Floyd–Warshall algorithm is given by the calculation of all the shortest paths among all pairs (i, j), and for all k. Let $n = |V|$; all the shortest paths are n^2, considering all pairs. The calculation of all $shortestPath(i, j, k)$ from $shortestPath(i, j, k-1)$ requires $2n^2$ operations. Recursively, to calculate all the paths from $k = 0$ requires n calculation, giving to the cost $n * 2n^2 = 2n^2 = \theta(n^3)$.

3.4 Power graph analysis

Often real networks have a complex structure (see for details in next chapter, Chapter 4). Therefore, there is the need to introduce methods that are able to provide a simpler representation. In bioinformatics and computational biology, the so-called **power graph analysis** is a common method for representing complex and large networks. At a glance, power graph analysis enables the transformation of a network into a **power representation** that is able to provide a compact representation and support for the effective visualization of the large networks [34].

Power graph analysis provides a compact representation of a graph by representing star (Fig. 3.8), bicliques (Fig. 3.9) and clique (Fig. 3.10) structures as power node and power edges. Therefore, a power graph has both simple nodes and power nodes that are cliques, bicliques, and stars. In this way, power graph analysis offers a lossless compression for graphs, since small regions of the graph are represented by using one of the compact representations provided by power graph analysis.

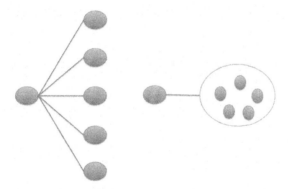

Figure 3.8. A star in a graph is a node connected with a set of nodes outside. In a power graph, a star is represented by a power edge between a regular node and a power node.

Formally, given a graph $\mathcal{G}$, its power graphs are defined on the power set of nodes as well as on the power set of edges of $\mathcal{G}$. In mathematics, a power set, $\mathcal{P}$, of a set I is a set of all subsets of I. Therefore, given a graph $\mathcal{G} = (\mathcal{V}, \mathcal{E})$, where $\mathcal{V} = \{v_0, \ldots, v_n\}$ is the set of nodes and $\mathcal{E} \subseteq \mathcal{V} \times \mathcal{V}$ is the set of edges; a power graph $\mathcal{G}' = (\mathcal{V}', \mathcal{E}')$ is a graph defined on the power set $\mathcal{V}' \subseteq \mathcal{P}(\mathcal{V})$ of power nodes connected to each other by power edges: $\mathcal{E}' \subseteq \mathcal{V}' \times \mathcal{V}'$. For each power graph, following conditions must hold.

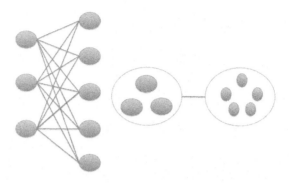

Figure 3.9. Bicliques are the two disjoint sets of nodes with an edge between every member of one set and every member of the other set. Each set is represented by a power node, and the biclique is represented as an edge between connecting them.

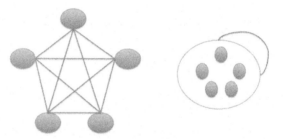

Figure 3.10. Clique is a set of nodes that are completely connected. In a power graph, a clique is represented by a power node with a loop.

- Given two power nodes, N_{p1} and N_{p2}, these will be connected by a power edge E_{p12}, if all nodes of the first power node are connected to all nodes of the second power node (star and biclique).
- A power node N_{p1} will be connected to itself by a power edge E_{p11} if all nodes in the power node are connected to each other by edges (clique).
- Any two power nodes are either disjoint, or one is included in the other.

Since power graph is a representation of an initial graph, it is interesting to have a minimal and unique representation of the input graph. Minimal power graph is defined as a power graph having the least number of power edges and nodes. Unfortunately, there is no unique minimal power graph possible for a given graph.

Finding a power graph is a two-step process (Fig. 3.11). The first step identifies the candidate power nodes through hierarchical clustering of the nodes in the network based on the similarity of their neighboring nodes. The similarity of the two sets of neighbors is taken as the Jaccard index of the two sets. Power edges that may replace the most edges in the original network are added first to the power graph.

---

procedure Power Graph Analysis

1. Hierarchical clustering of the nodes: This finds candidate power nodes by a hierarchical clustering of the nodes in the network.

2. Greedy search for possible power edges: Greedily possible power edges are searched between the power nodes to form the power graph.

Figure 3.11. Power graph analysis steps.

While analyzing any network, sometimes it is necessary to identify crucial or vital nodes. In a biological network, it is important to discover any or set of key genes playing major regulatory roles in diseased environment [11,25]. Elucidating such key genes may help in developing small molecules that bind those key genes. Finding an influential node or a person within a social graph is another prominent application domain of node centrality analysis. Next, we discuss various centrality measures commonly used in real-life graph analysis.

3.5 Network centrality measures

As discussed before, a graph or a network of n ($n = |\mathcal{V}|$) nodes can be represented as an adjacency matrix (($\mathcal{A} \in \Re^{n \times n}$), where each entry in the matrix $\mathcal{A}_{ij} \neq 0$ indicates the existence of an edge between nodes i and j, and $\mathcal{A}_{ij} = 0$ indicates the absence of an edge between the two nodes. A special case of graphs, called **edge-weighted graphs**, is characterized by a special adjacency matrix, containing real-valued numbers included in the interval $(0, 1)$. The following discussion is focused on undirected nonweighted graphs, and may be easily extended to both ordered and edge-weighted graphs.

In the case of an unweighted network, a *geodesic* (or shortest path) from node w_i to node w_j is the path that involves the minimum number of edges. Consequently, we may define the *distance* between nodes w_i and w_j, where $\rho_{\mathbf{g}}(w_i, w_j)$ is the number of edges involved in a geodesic between w_i and w_j. Starting from the computation of distance, a set of *centrality* measures have been introduced. The aim of such measure is to evidence the relevance, or importance, of a node in a network by analyzing the topology.

3.5.1 Degree centrality

Given a node, w_i, the **degree centrality** is the number of adjacent nodes to w_i, which can be defined as follows:

$$C_{deg}(w_i) = deg(w_i).$$

Sometimes, the degree centrality is normalized by the maximal possible degree of a node to obtain a number between 0 and 1:

$$C_{degnorm}(w_i) = deg(w_i)\frac{d_i(\mathbf{g})}{n-1}.$$

Degree centrality is an obvious measure, and its computation is also simple. It gives some information related to the relevance of the node w_i, but it misses some relevant aspects of the whole structure of the network as well as the node's position. In addition to degree centrality, the other centrality measures are based on the calculation of the geodesic paths; the analysis of the distance of a node concerning the others is discussed next.

3.5.2 Closeness centralities

The rationale of the **closeness centrality** is to consider a central node that is *close* to the others in terms of distance. Formally, the closeness centrality of a node w_i is the reciprocal of the average shortest path distance to w_i over all $n-1$ reachable nodes, i.e.,

$$C_{closeness}(w_i) = \frac{n-1}{\sum_{j=1}^{j=n-1, j\neq i} d(w_i, w_j)},$$

where $d(w_i, w_j)$ is the shortest distance between w_i and w_j. Closeness centrality has a simple calculation and it measures how a node is reachable by the others. In such formulation, each connected component of a graph presents a different closeness centrality distribution. In order to overcome this limitation, Wasserman and Faust proposed an improved formula for graphs with more than one connected components [36].

3.5.3 Betweenness centralities

The closeness centrality indicates how a node is close to the other, whereas sometimes it is important to evaluate how a node stands between each other. For these aims the **betweenness centrality** has been introduced [7]. Betweenness centrality first calculates all the shortest paths among all node pairs. Then, for each node w_i, the number of shortest paths that pass through w_i is calculated, and each node receives a score based on the number of these shortest paths that pass through. Consequently, nodes that are included most frequently into these shortest paths will get higher betweenness centrality scores. Formally, the betweenness centrality of a node (w_i), can be expressed as follows.

$$C_{betweennes}(w_i) = \sum_{i \neq j \neq k} \frac{\sigma_{j,k}(i)}{\sigma_{j,k}},$$

where, $\sigma_{j,k}$ is the total number of shortest paths from node w_j to node w_k, and $\sigma_{j,k}(i)$ is the number of those paths that pass through i.

3.5.4 Eigenvector centralities

Eigenvector centrality is proposed by [6], is a measure of the influence of a node in a network. It scores all nodes of a network on the assumption that connections to high-scoring nodes contribute more to the score of the node rather than connections to low-scoring nodes. Interestingly, Google's page ranking algorithm is a variant of eigenvector centrality measure.

Given an adjacency matrix $\mathcal{A}$, the relative centrality score a node v can be defined as

$$x_v = \frac{1}{\lambda} \sum_{w \in Neigh(v)} x_w = \frac{1}{\lambda} \sum_{w \in \mathcal{G}} \mathcal{A}_{v,w} x_w.$$

Let x_i be the eigenvector centrality of the ith node v_i, then $X = (x_1, x_2, \cdots x_n)^T$, the solution of equation $AX = \lambda X$, where λ is the greatest eigenvalue of $\mathcal{A}$ to ensure that all values x_i are positive [29] by the Perron–Frobenius theorem. The vth component of the related eigenvector will give the relative centrality score of the vertex v in the network.

3.6 Graph community and discovery

Given a real graph, the distribution of edges, as noted in [18,21], is quite inhomogeneous, both globally and locally. This causes the insurgence of regions with a high concentration of edges, i.e., subsets of nodes with a high number of edges among them, and a low number of edges outside. These regions are often referred to as *communities, (or clusters or modules),* and the property of graphs that have these regions is called a *community structure*. The relevance of finding communities in a graph is high with applications in many fields. There exist many algorithms for community detection in graphs, and the complete enumeration of them is beyond the scope of this book. Here, we define the main characteristics of a community, then we present some approaches, whereas the interested reader may find more information in [18, 21].

As outlined by Fortunato in [18], there is no universal definition of community, and the existing ones are often related to the specific application or context. Trivially, a community in a graph should have two main properties: the number of edges connecting its nodes should be *high*, whereas the number of the edges connecting its edges to the remaining ones should be *low*.

More specifically, a community is popularly defined as a cluster having significant intra-community connectivity in comparison to inter-community connectivity in the network. Formally, we can define a network community as follows:

Definition 3.6.1 (Network community). Given a network $\mathcal{G} = (\mathcal{V}, \mathcal{E})$, a network community $C_i = (\mathcal{V}', \mathcal{E}')$ is a densely connected subgraph of $\mathcal{G}$ ($C_i \subseteq \mathcal{G}$), where interconnectivity of $\mathcal{V}'$ with respect to $\mathcal{E}' \subseteq \mathcal{E}$ is higher in comparison to the rest of $\mathcal{V}$, i.e., $\mathcal{V} - \mathcal{V}'$.

Consequently, a community should have three desired properties:
- the intra-community density is significantly larger than the average edge density of the graph;
- the inter-community density is significantly lower than the average edge density of the graph;
- C_i should be connected, i.e., there must be a path between each pair of its vertices, running only through vertices of C_i.

Mostly, communities are treated as exclusive, also called disjoint community. An exclusive community is a subgraph, such that none of its vertices belongs to any other communities. In other words, a node belonging to a community cannot be a member of any other communities simultaneously.

Definition 3.6.2 (Disjoint community). Given a set of k communities $C = \{C_1, C_2, \cdots C_k\}$, derived from $\mathcal{G}$, a community C_i is a disjoint community (also called exclusive community) if no communities share common members, i.e., $C_i \cap C_j = \phi, \forall i, j = 1, \cdots k$ and $i \neq j$.

In real life, it is unrealistic to consider communities as exclusive. In the context of the social network, a member may exhibit an association with two different interest groups simultaneously. Hence, traditional clustering is not applicable to form communities from the overlapping distribution of the interests. As opposed to disjoint communities, overlapping communities allow members sharing membership with different communities.

Definition 3.6.3 (Overlapping communities). Two communities $C_i = (V_i, \mathcal{E}_i)$ and $C_j = (V_j, \mathcal{E}_j)$, where C_i and $C_j \in C$ are overlapping if $V_i \cap V_j \neq \phi$.

A more special case of an overlapping community may persist in the real networks, where some of the members within a community possesses similar characteristics, or inclined towards common interest groups different from the rest of the members of the community. It leads to the formation of a more compact community within a community, termed as an intrinsic [27] or embedded community.

Definition 3.6.4 (Intrinsic community). A community C_i is embedded or intrinsic within C_j if $C_i \subset C_j$ and the connectedness density of nodes within C_i, $\rho(C_i)$ is significantly different from the density of C_j, $\rho(C_j)$.

$$Intrinsic(C_i, C_j) = \begin{cases} 1, & \text{if } C_i \subset C_j \\ & \text{and } |\rho(C_i) - \rho(C_j)| > \xi \\ 0, & \text{otherwise}, \end{cases} \qquad (3.2)$$

where ξ is a user defined significant density difference threshold.

A pictorial illustration of all three types of communities discussed is given in Fig. 3.12 for better understanding. Unlike an intrinsic community, which may survive as an independent community (Community B), a member belonging to overlapping communities does not have any independent identity without the participating communities (say, Community A and B).

In biological networks, existence of overlapping and intrinsic communities (popularly known as network modules or motif) has been established in different researches [25]. More detail discussion on biological network modules are available in Chapter 6 and 7.

Identification of compact communities in a given network depends on the computational methodology employed to extract the same. Next, we report different community detection methods available.

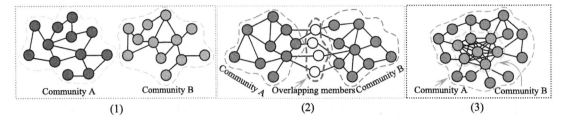

Figure 3.12. Representation of disjoint (1), overlapping (2) and intrinsic (3) communities [27]. (1) Denotes the presence of two disjoint communities. Yellow (light gray in print version) color nodes in (2) are the members showing common interests in both the communities. In (3) Community B is embedded inside the community A.

3.6.1 Few community detection methods

Infomap [32] is a popular community detection method for social networks that employ random walks in the connected network to maximize the amount of information flow on the decomposed subnetworks for community detection. **iDBLINK** [26] is an incremental density-based community detection algorithm [33] for dynamic network, where the structure of the communities changes over time. Based on the changes in the connections between nodes over time, it immediately updates the corresponding community structures. Inspired from fuzzy granular social networks (**FGSN**) [20] and Louvain algorithm [5], Nicole et. al. [13] proposed a method for detecting communities in a social network. The molecular complex detection algorithm (**MCODE**),[1] described in the early work of Bader et al., [30], takes in input and interaction networks and tries to find complexes by building clusters. The rationale of MCODE is the separation of dense regions based on an ad hoc defined local density. MCODE has three main stages: (i) node weighting, (ii) complexes prediction, and (iii) post-processing.

Overlapping cluster generator (**OCG**) [2] creates a hierarchy of overlapping clusters according to an extension of Newman's modularity function. Preferential learning and label propagation algorithm, called **PLPA** [35] is proposed to detect overlapping communities based on learning behavior and information interaction in social networks. **COPRA** [19] optimizes and expands seed community iteratively, based on the clustering coefficient of each node by taking the average clustering coefficient of neighboring nodes. **LPANNI** [23], is a label propagation-based method to detect overlapping communities. It quantifies the influence of neighbor nodes by adopting node importance and affinity between node pairs to improve the label update strategy. **SLPAD** [1]

---

[1]http://baderlab.org/Software/MCODE.

follows label propagation model, such as SLPA [37], for detecting overlapping communities in dynamic networks. **AFOCS** [28] is an adaptive framework for the detection of overlapping communities as well as tracking the evolution of overlapping communities in incremental networks. **TILES** [31] extracts overlapping communities in dynamic social networks using peripheral membership and core membership and reevaluate membership of nodes to communities for each new interaction. **OSLOM** [22] uses local optimization of a fitness function to detect overlapping communities in incremental networks. An extended adaptive density peaks clustering (**EADP**) [38] is proposed for overlapping community detection based on a novel distance function. **EADP** adaptively choose cluster centres by employing a linear fitting-based strategy. **MCL** (Markov clustering algorithm) [14] is an iterative algorithm that simulates random walks using Markov chains. A possible way to define a module within a network is as a collection of nodes that are more connected with each other than to the others. It follows that a random walk starting in any of these nodes is more likely to stay within the cluster rather than to travel between clusters. The simulation is performed by iteratively applying two main operations, usually referred to as expansion and inflation.

Next, we demonstrate the use of R for the implementation of various graph analysis methods.

3.7 R scripts for graph analysis

R scripting language is rich with various graph analysis packages. We demonstrated the use of *igraph* library in Chapter 2. We extend our demonstration for the use of igrah library in implementing few of the graph analysis algorithm as discussed above. It starts with installing the library as demonstrated in the code given in Fig. 2.18. We implement graph traversal algorithms at first. Various community detection algorithms and their outcomes are also visualized (Figs. 3.13, 3.14, 3.15, 3.16). We also demonstrate how to use R code for implementing different centrality measures. For more codes one can refer igraph manual.[2]

---

[2]https://cran.r-project.org/web/packages/igraph/igraph.pdf.

3.7.1 Graph traversal using R

Breadth-First Search (BFS)

```r
net= graph.tree(50, children=8, mode="undirected" )
bfs(net, root=1, "out", order=TRUE, rank=TRUE, father=
    TRUE, pred=TRUE, succ=TRUE, dist=TRUE)

f=function(graph, data, extra) {
  print(data)
  FALSE
  } // Use a callback
tmp=bfs(net, root=1, "out", callback=f)

## How to use a callback to stop the search
## We stop after visiting all vertices in the initial
    component

f=function(graph, data, extra) {
  data['succ'] == -1
  } // Stop the search
bfs(net, root=1, callback=f)
```

Depth-First Search (DFS)

```r
net=graph.tree(50, children=8, mode="undirected" )
dfs(net, root=1, "out", TRUE, TRUE, TRUE, TRUE)

f.in=function(graph, data, extra) {
  cat("in:", paste(collapse=", ", data), "\n")
  FALSE
  }
f.out=function(graph, data, extra) {
  cat("out:", paste(collapse=", ", data), "\n")
  FALSE
  }

tmp=dfs(net, root=1, "out", in.callback=f.in, out.
    callback=f.out) // Use a callback

f.out=function(graph, data, extra) {
  data['vid'] == 1
  } // Terminate after the first component
tmp=dfs(net, root=1, out.callback=f.out)
```

Dijkstra's Shortest Path

```
distance_table(graph, directed=TRUE) // The graph to work
    on

mean_distance(graph, directed=TRUE, unconnected=TRUE)

distances(graph, v=V(graph), to=V(graph), mode=c("all", "
    out", "in"), weights=NULL, algorithm="dijkstra")

shortest_paths(graph, from, to=V(graph), mode=c("out", "
    all", "in"), weights=NULL, output=c("vpath", "epath
    ", "both"), predecessors=FALSE, inbound.edges=FALSE)

all_shortest_paths(graph, from, to=V(graph), mode=c("out
    ", "all", "in"),  weights=NULL)
```

Floyd-Warshall Algorithm

```
floyd(x) // x -> The adjacency matrix of a directed graph
```

or

Install Package **e1071**

```
allShortestPaths(x) // x -> matrix or distance object
extractPath(obj, start, end) // obj -> return value of
    allShortestPaths, start -> (integer) starting point
    of path, end -> (integer) end point of path
```

3.7.2 R scripts for community discovery

Edge betweenness based (Newman-Girvan)

```
net=graph.tree(40, children=5, mode="undirected"
club=cluster_edge_betweenness(net)
dendPlot(club, mode="hclust")
plot(club, net)
```

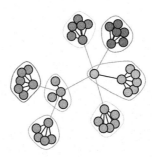

Figure 3.13. Community finding based on edge betweenness.

Label propagation based

```
clup=cluster_label_prop(net) // net -> Zachary's karate
    club
plot(clup, net)
```

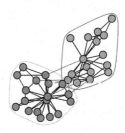

Figure 3.14. Community finding based on label propagation.

Modularity optimization

```
clug=cluster_fast_greedy(as.undirected(net))
plot(clug, as.undirected(net))
```

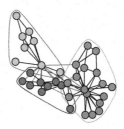

Figure 3.15. Community finding based on modularity optimization.

k-mean clustering

```
kmeans_model=kmeans(x=X, centers=m) // X is the data
    matrix and m is the number of clusters.
```

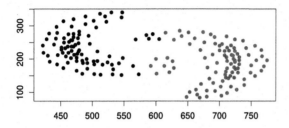

Figure 3.16. k-mean clustering.

Fuzzy c-mean clustering

Install the package **e1071**

```
fclus=cmeans(x, centers, dist="euclidean", method="cmeans
    ") // x = The data matrix, centers = Number of
    clusters, dist = Must be one of the following: If "
    euclidean", the mean square error, if "manhattan",
    the mean absolute error is computed, method = If "
    cmeans", then we have the cmeans fuzzy clustering
    method.
```

3.7.3 Network centrality analysis in R

Degree Centrality

```
z=40
net=barabasi.game(z) // choose your network
deg=degree(net, mode="all")
hist(deg)
```

Closeness Centrality

```
z=150
net=erdos.renyi.game(z, 0.1)
closeness(net)
```

```
closeness(net, mode="in") // mode -> Character string,
   defined the types of the paths used for measuring
   the distance in directed graphs. "in" measures the
   paths to a vertex, "out" measures paths from a
   vertex, "all" uses undirected paths. This argument
   is ignored for undirected graphs.
closeness(net, mode="out")
closeness(net, mode="all")
```

Betweenness Centrality

```
z=150
net=erdos.renyi.game(z, 0.1)
betweenness(net)
edge_betweenness(net)
```

Eigenvector Centrality

```
g=make_ring(10, directed=FALSE) // Generate test data
eigen_centrality(g) // Compute eigenvector centrality
   scores
```

3.8 Summary

We presented some of the main algorithms for graph analysis. We started by discussing the approaches for graph traversal, and successively we showed how these algorithms are the basis for graph analysis. Node centrality measures are highlighted with R code implementations. We also considered the compression of graphs through power graph analysis and the discovery of communities.

We next discuss various complex network models and their properties, which do not follow conventional or regular graph properties.

Acknowledgment

Authors thank Dr. Keshab Nath for his contribution while writing this chapter.

References

1. Nathan Aston, Jacob Hertzler, Wei Hu, et al., Overlapping community detection in dynamic networks, Journal of Software Engineering and Applications 7 (10) (2014) 872.
2. Emmanuelle Becker, Benoît Robisson, Charles E. Chapple, Alain Guénoche, Christine Brun, Multifunctional proteins revealed by overlapping clustering in protein interaction network, Bioinformatics 28 (1) (2012) 84–90.
3. Richard Bellman, On a routing problem, Quarterly of Applied Mathematics 16 (1) (1958) 87–90.
4. Richard Bellman, Dynamic programming, Science 153 (3731) (1966) 34–37.
5. Vincent D. Blondel, Jean-Loup Guillaume, Renaud Lambiotte, Etienne Lefebvre, Fast unfolding of communities in large networks, Journal of Statistical Mechanics: Theory and Experiment 2008 (10) (2008) P10008.
6. Phillip Bonacich, Some unique properties of eigenvector centrality, Social Networks 29 (4) (2007) 555–564.
7. Ulrik Brandes, Christian Pich, Centrality estimation in large networks, International Journal of Bifurcation and Chaos 17 (07) (2007) 2303–2318.
8. C.J. Cheney, A nonrecursive list compacting algorithm, Communications of the ACM 13 (11) (November 1970) 677–678.
9. Boris V. Cherkassky, Andrew V. Goldberg, Tomasz Radzik, Shortest paths algorithms: theory and experimental evaluation, Mathematical Programming 73 (2) (1996) 129–174.
10. Thomas H. Cormen, Charles E. Leiserson, Ronald L. Rivest, Clifford Stein, Introduction to Algorithms, MIT Press, 2009.
11. Nick Dand, Reiner Schulz, Michael E. Weale, Laura Southgate, Rebecca J. Oakey, Michael A. Simpson, Thomas Schlitt, Network-informed gene ranking tackles genetic heterogeneity in exome-sequencing studies of monogenic disease, Human Mutation 36 (12) (2015) 1135–1144.
12. Edsger W. Dijkstra, A note on two problems in connexion with graphs, Numerische Mathematik 1 (1) (1959) 269–271.
13. Nicole Belinda Dillen, Aruna Chakraborty, Modularity-based community detection in fuzzy granular social networks, in: Proceedings of the International Congress on Information and Communication Technology, Springer, 2016, pp. 577–585.
14. A.J. Enright, S. Van Dongen, C.A. Ouzounis, An efficient algorithm for large-scale detection of protein families, Nucleic Acids Research 30 (7) (April 2002) 1575–1584.
15. Robert W. Floyd, Algorithm 97: shortest path, Communications of the ACM 5 (6) (1962) 345.
16. L.R. Ford, D.R. Fulkerson, Maximal flow through a network, Canadian Journal of Mathematics 8 (1956) 399–404.
17. Lester R. Ford Jr., Network flow theory, Technical report, Rand Corp Santa Monica Ca, 1956.
18. Santo Fortunato, Community detection in graphs, Physics Reports 486 (3–5) (2010) 75–174.
19. Steve Gregory, Finding overlapping communities in networks by label propagation, New Journal of Physics 12 (10) (2010) 103018.
20. Kundu Suman, Sankar K. Pal, Fuzzy-rough community in social networks, Pattern Recognition Letters 67 (2015) 145–152.

21. Andrea Lancichinetti, Santo Fortunato, Benchmarks for testing community detection algorithms on directed and weighted graphs with overlapping communities, Physical Review E 80 (1) (2009) 016118.

22. Andrea Lancichinetti, Filippo Radicchi, José J. Ramasco, Santo Fortunato, Finding statistically significant communities in networks, PLoS ONE 6 (4) (2011) e18961.

23. Meilian Lu, Zhenlin Zhang, Zhihe Qu, Yu Kang, Lpanni: overlapping community detection using label propagation in large-scale complex networks, IEEE Transactions on Knowledge and Data Engineering 31 (9) (2018) 1736–1749.

24. Andrew Lumsdaine, Douglas Gregor, Bruce Hendrickson, Jonathan Berry, Challenges in parallel graph processing, Parallel Processing Letters 17 (01) (2007) 5–20.

25. Hazel Nicolette Manners, Swarup Roy, Jugal K. Kalita, Intrinsic-overlapping co-expression module detection with application to Alzheimer's disease, Computational Biology and Chemistry 77 (2018) 373–389.

26. Fanrong Meng, Feng Zhang, Mu Zhu, Yan Xing, Zhixiao Wang, Jihong Shi, Incremental density-based link clustering algorithm for community detection in dynamic networks, Mathematical Problems in Engineering 2016 (2016).

27. Keshab Nath, Swarup Roy, Detecting intrinsic communities in evolving networks, Social Network Analysis and Mining 9 (1) (2019) 13.

28. Nam P. Nguyen, Thang N. Dinh, Sindhura Tokala, My T. Thai, Overlapping communities in dynamic networks: their detection and mobile applications, in: Proceedings of the 17th Annual International Conference on Mobile Computing and Networking, ACM, 2011, pp. 85–96.

29. Kazuya Okamoto, Wei Chen, Xiang-Yang Li, Ranking of closeness centrality for large-scale social networks, in: International Workshop on Frontiers in Algorithmics, Springer, 2008, pp. 186–195.

30. Yanjun Qi, Fernanda Balem, Christos Faloutsos, Judith Klein-Seetharaman, Ziv Bar-Joseph, Protein complex identification by supervised graph local clustering, Bioinformatics 24 (13) (2008) i250–i268.

31. Giulio Rossetti, Luca Pappalardo, Dino Pedreschi, Fosca Giannotti, Tiles: an online algorithm for community discovery in dynamic social networks, Machine Learning 106 (8) (2017) 1213–1241.

32. Martin Rosvall, Carl T. Bergstrom, Maps of random walks on complex networks reveal community structure, Proceedings of the National Academy of Sciences 105 (4) (2008) 1118–1123.

33. Swarup Roy, D.K. Bhattacharyya, An approach to find embedded clusters using density based techniques, in: Distributed Computing and Internet Technology, Springer, 2005, pp. 523–535.

34. Loic Royer, Matthias Reimann, Bill Andreopoulos, Michael Schroeder, Unraveling protein networks with power graph analysis, PLoS Computational Biology 4 (7) (2008) e1000108.

35. JinFang Sheng, Kai Wang, ZeJun Sun, Bin Wang, FaizaRiaz Khawaja, Ben Lu, JunKai Zhang, Overlapping community detection via preferential learning model, Physica A: Statistical Mechanics and Its Applications 527 (2019) 121265.

36. Stanley Wasserman, Katherine Faust, Social Network Analysis: Methods and Applications, vol. 8, Cambridge University Press, 1994.

37. Jierui Xie, Boleslaw K. Szymanski, Xiaoming Liu, Slpa: uncovering overlapping communities in social networks via a speaker-listener

interaction dynamic process, in: Data Mining Workshops (ICDMW), 2011 IEEE 11th International Conference on, IEEE, 2011, pp. 344–349.

38. Mingli Xu, Yuhua Li, Ruixuan Li, Fuhao Zou, Xiwu Gu, Eadp: an extended adaptive density peaks clustering for overlapping community detection in social networks, Neurocomputing 337 (2019) 287–302.

4

Complex network models

Contents

To understand various biological networks, it is essential to appreciate the underlying topological structures or models of the network. In this chapter, we first discuss various complex network properties, followed by an introduction to different network models. We also highlight use of R packages available for network analysis.

4.1 Complex graphs

Over the past few decades, large real-world systems, such as WWW, Internet, social networks, wireless networks, supply-chain networks, are often modeled as a graph for the ease of computational analysis. The interaction relationships across different entities or macromolecules in the biological systems, such as protein-

Biological Network Analysis. https://doi.org/10.1016/B978-0-12-819350-1.00010-4

protein interactions (PPI), gene regulation and association, signaling pathways, metabolic activities, neuronal connectivity of a brain, etc., are also modeled as graphs or networks. However, such graphs are not simple graphs as regular graphs. Due to their unconventional topological properties, they are often treated as **complex graphs or networks**. Unlike conventional graphs, real-world networks exhibit non-trivial topological properties, such as varying degree distributions, high or low clustering coefficients, degree assortativity or average path length of the network. Accordingly, networks are classified into different models.

Before discussing various network models, it is important to understand the topological characteristics of any complex network. Usually, network models are defined based on such characteristics only. Thus we discuss topological characteristics first before introducing available models.

4.2 Topological characteristics of networks

In a network, the interconnection patterns among the nodes are termed as *network topology*. The varying topological properties of any complex networks make the task of network comparison and classification a challenging activity. Therefore a set of summary statistics or quantitative performance measures are important to describe and compare the complex networks. In the last few years, many quantities and measures are proposed and investigated for complex network analysis. However, among all, three measures, namely *average path length* (L) [2], *clustering coefficient* (CC) [16,33], and *degree distribution* (P_k) [1,3] play a key role in complex network analysis. Next, we discuss different topological characteristics considered for any complex networks.

4.2.1 Average path length

Average path length (L) is one of the most robust assessment measures for network topology study. It quantifies how complex real-world networks are "wired" and evolving. Moreover, the average path length is a measure of network size, and it indicates the rate of (quick) transfer of information throughout the network. Average path length of a network $\mathcal{G}(\mathcal{V}, \mathcal{E})$, is the mean distance of all possible shortest path ($d_{i,j}$) followed between any two nodes, v_i and v_j. Mathematically, an average path length for a directed graph can be expressed as

$$L = \frac{1}{|\mathcal{V}|(|\mathcal{V}|-1)} \sum_{i=1}^{|\mathcal{V}|} \sum_{j=1}^{|\mathcal{V}|} d_{i,j}, \qquad (4.1)$$

where, $d_{i,j}$ is the shortest path between any two nodes, i and j, and $|\mathcal{V}|(|\mathcal{V}|-1)$ is the total number of expected edges. In the case of an undirected graph, where $e(i,j) = e(j,i)\ \forall e(i,j) \in \mathcal{E}$, the average path length for an undirected graph can be represent as

$$L = \frac{2}{|\mathcal{V}|(|\mathcal{V}|-1)} \sum_{i=1}^{|\mathcal{V}|} \sum_{j=1}^{|\mathcal{V}|} d_{i,j}. \qquad (4.2)$$

Most of the real-world networks have a small average path length, where every node is connected through the shortest path to every other nodes. With the change in the number of nodes in a network, the average path length is also affected, but that change is not drastic.

4.2.2 Clustering coefficient

Clustering coefficient (CC) is a measure of affinity (likelihood), to which nodes in a network tends to create tightly connected group with each others. The tendency of likelihood of adjacent nodes in a network is higher in comparison to the nonadjacent nodes. There exists several alternatives [20], [10], [30] for defining clustering coefficient. Clemente et al. [5] generalized clustering coefficient measure for weighted and directed networks. Latapy et al. [19] and Opsahl [24] defined a new clustering coefficient measure for bipartite graph. However, among all, Watts and Strogatz [33] definitions of clustering coefficient is widely accepted. Furthermore, they introduced the concept of *local* and *global*, or network *average* clustering coefficient in their proposed approach. Local clustering coefficient (CC_{v_i}) is the ratio of total number of edges that are present among the neighbors of a node v_i to the total number of possible edges that could exist among the neighbors of v_i. Thus CC_{v_i} for a directed graph is given as

$$CC_{v_i} = \frac{\sum_{j=1}^{|N_{v_i}|} \lambda(v_i, v_j)}{|N_{v_i}|(|N_{v_i}|-1)}, \qquad (4.3)$$

where, $\lambda(v_i, v_j) = \begin{cases} 1, & \text{if } (v_i, v_j) \text{ is connected, } \forall v_j \in N_{v_i}, i \neq j \\ 0, & \text{otherwise.} \end{cases}$

$N_{v_i} = \{v_k | e(i,k) \in \mathcal{E} \vee e(k,i) \in \mathcal{E}\}$ is the set of adjacent nodes of v_i in $\mathcal{V}$, and $|N_{v_i}|(|N_{v_i}| - 1)$ is the total number of expected edges.

In the case of an undirected graph, the total number of expected edges will be $\frac{|N_{v_i}|(|N_{v_i}|-1)}{2}$, since $e(i,j) = e(j,i)$. Thus CC_{v_i} for an undirected graph can be represent as

$$CC_{v_i} = \frac{2 \times \sum_{j=1}^{|N_{v_i}|} \lambda(v_i, v_j)}{|N_{v_i}|(|N_{v_i}| - 1)}. \qquad (4.4)$$

The average (global) clustering coefficient [33] is the mean of K local clustering coefficient. Therefore the global clustering coefficient for a graph $\mathcal{G}$ can be defined as

$$\overline{CC} = \frac{\sum_{i=1}^{\mathcal{V}} CC_{v_i}}{\mathcal{V}}, \qquad (4.5)$$

where the range of $\overline{CC}$ values lies within $0 \leq \overline{CC} \leq 1$.

According to Luce and Perry [20], the global clustering coefficient can be defined as the ratio of the total number of closed triplets (or 3 × triangles) to the total number of all triplets (both open and closed) (Fig. 4.1). An open triplet consists of three nodes connected by two undirected ties. Whereas, in a closed triplet, all three nodes are connected by three undirected ties. Therefore a triangle consists of three closed triplets. Thus mathematically global clustering coefficient can be defined as

$$CC = \frac{Total\ number\ of\ closed\ triplets\ (or\ 3 \times triangles)}{Total\ number\ of\ all\ triplets}. \qquad (4.6)$$

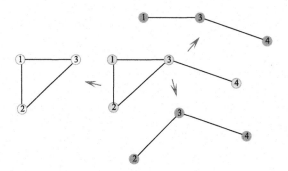

Figure 4.1. Example shows the *closed triplets* [(1,2,3), (1,3,2), (2,1,3)] (yellow [light gray in print version] color) and the *open triplets* [(1,3,4), (2,3,4)] (red [gray in print version] color) in a network [(1, 2, 3, 4)] (blue [dark gray in print version]).

Fig. 4.2 represents the clustering coefficient of three real-world networks, namely email, Gnutella P2P, and Youtube network.

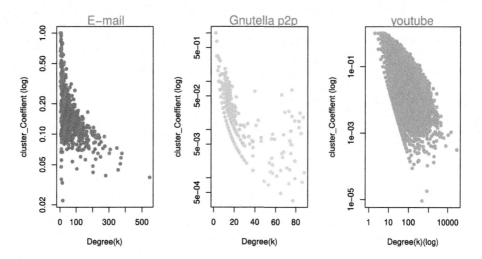

Figure 4.2. Clustering coefficient of few real world networks [4]. Local clustering coefficient values (y-axis) are plotted against nodes with k degree (x-axis). For larger network (youtube) with large range of k values log values have been used to contain within limited axis margin.

4.2.3 Degree distribution

Degree distributions (P_k) is the probability that a node chosen randomly has a degree k. Therefore, P_k of a network is the fraction of nodes (n) having degree $k(n_k)$. Assume a network with n number of nodes and n_k numbers of nodes having degree k, then, P_k can be defined as follows:

$$P_k = \frac{n_k}{n}. \tag{4.7}$$

For a large network with $\mathcal{V}$ nodes (where average degree $\langle k \rangle \ll |\mathcal{V}|$), the degree distribution (4.8) approximately follows the Poisson distribution:

$$P_k = e^{-\langle k \rangle} \frac{\langle k \rangle^k}{k!}. \tag{4.8}$$

Fig. 4.3 depicts the degree distribution of real-world networks, namely *facebook*, *ca-GrQc*, and *ca-HepTh* network. The degree distribution of real-world networks (like the Internet, social network, etc.) found to follow the *power law* (functional relationship) properties, which is defined as follows:

$$P_k \sim k^{-\gamma}, \tag{4.9}$$

where γ is a constant, and its value is bounded between 2 and 3.

Such a pattern is called a *power law distribution*, or a *scale-free distribution*, because the shape of the distribution does not change with scale (*see* Fig. 4.7).

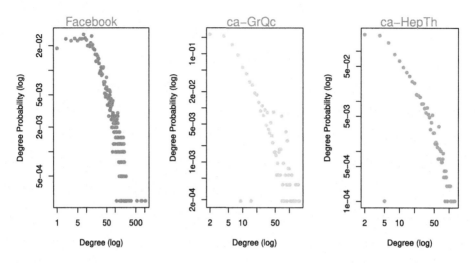

Figure 4.3. Degree distribution of real-world networks [4]. Probability (y-axis) of nodes with degree k (x-axis) is plotted. For larger network with large range of k values and probability scores, log values have been used to contain within limited axis margin.

4.2.4 Rich club coefficient

The rich-club coefficient, introduced by Zhou and Mondragon in the context of the Internet topology [36], refers to the tendency of high-degree nodes (i.e., the hubs) in the network, to be very well-connected to other hub nodes. The name "*rich-club*" arises from the metaphor that the nodes with a large number of links, i.e., the hubs are "*rich*", and they tend to be tightly and well-interconnected between themselves, forming subgraphs called "*club*". The rich-club coefficient is nothing but the measure of connectedness density within the *club*. A network with a rich club organization is shown in Fig. 4.4 for better understanding.

The nodes in a network can be categorized by a ranking scheme [36] or by their degree [8]. The rank r of a node represents the corresponding position of the node in the list of descending order of node degrees, i.e., the most highly-connected node is ranked as $r = 1$, the second best-connected node is $r = 2$, and so on. The density of connections between the r richest nodes is evaluated by the rich-club coefficient [36],

$$\Phi(r) = \frac{2E(r)}{r(r-1)}, \tag{4.10}$$

where $E(r)$ is the total number of links between r hub nodes and $r(r-1)/2$ is the maximum possible number of links among these nodes. Similarly, the rich-club coefficient [8] in terms of node de-

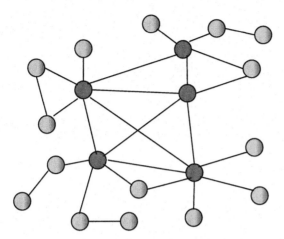

Figure 4.4. Hub nodes in red (gray in print version) colors are connected strongly, forming rich club of hubs.

gree can be represented as follows:

$$\Phi(k) = \frac{2E_k}{N_k(N_k - 1)}, \tag{4.11}$$

where E_k is the number of links present between the nodes of degree greater than or equal to k, and N_k is the number of such nodes. Therefore, $\Phi(k)$ measures the fraction of actual links connecting those nodes and the maximum number of possible links. This measure explicitly reflects how densely connected are the nodes within a network.

The behavior of the rich-club coefficient is proportional to the value of k. It means, a rich-club coefficient increasing with the degree k indicates that there exists a *rich-club* of nodes, which are densely interconnected than the nodes with smaller degrees. Contrarily, a decrease in the value of $\Phi(k)$ indicates the presence of many loosely connected and relatively independent subgroups. It is known as *rich-club phenomenon.*

4.2.5 Assortativity

Assortativity or *assortative mixing* was introduced by Newman [21]; is the tendency of nodes of a network (like social networks) to associate with others that are similar in some way. On the other hand, in nonsocial networks, such as biological networks, nodes with a high degree have a preference to associate with low-degree nodes. This tendency is referred as *disassortative mixing,* or *disassortativity.*

Assortativity is often quantified by the Pearson correlation between the *excess degree* distribution q_k and the *joint probability* distribution $e_{j,k}$ [21]. The excess degree is the number of edges leaving the node, other than the one that connects the pair. Similarly, the joint probability distribution is the distribution of the excess degrees of the two nodes at either end of a randomly chosen link. For an undirected graph, the assortativity is measured in terms of normalized Pearson coefficient of $e_{j,k}$ and q_k, and can be defined as

$$\rho = \frac{\sum_{jk} jk(e_{jk} - q_j q_k)}{\sigma_q^2}, \tag{4.12}$$

where, δ is the standard deviation the of remaining degree distribution and q_k is derived from the degree distribution P_k as

$$q_k = \frac{(k+1)P_{k+1}}{\sum_{j \geq 1} j P_j}. \tag{4.13}$$

In general, ρ has a range from -1 to 1, where 1 means a network has perfect assortativity, i.e., all nodes connect only with the nodes of a similar degree. If $\rho = 0$, then the network has no assortativity, which means any node can randomly connect to any other node. Whereas, at $\rho = -1$, the network is completely disassortative; all nodes connect with the nodes of different degrees.

For directed networks, assortativity is measured with four matrices, $\rho(in, in)$, $\rho(in, out)$, $\rho(out, in)$, and $\rho(out, out)$ [12,26], where *in* and *out* represent the probability distribution of links going *into* the target nodes and links going *out* of source nodes, respectively. For a given network of total number of edges $\mathcal{E} = \{1, 2, \cdots, n\}$, the $\rho(out, in)$ can be defined as follows:

$$\rho(out, in) = \frac{\sum_i (j_i^{out} - j^{\overline{out}})(k_i^{in} - k^{\overline{in}})}{\sqrt{\sum_i (j_i^{out} - j^{\overline{out}})^2} \sqrt{\sum_i (k_i^{in} - k^{\overline{in}})^2}}, \tag{4.14}$$

where, j_i^{out} and k_i^{in} are the *out*-degree of the source node and *in*-degree of the target node of an edge i, respectively. Similarly, $j^{\overline{out}}$ and $k^{\overline{in}}$ are the average *out*-degree of sources and *in*-degree of targets, respectively.

For a given network, assortativity is measured at the global level, where an individual node contribution to the network assortativity is missing. Moreover, there could be nodes that are disassortative, and vice-versa. M. Piraveenan et al. [27] proposed a local assortative measure, which can examine the assortative or disassortative behavior of each node at the local level. Local assortativity in undirected networks is defined as

$$\rho = \frac{j\ (j+1)\left(\overline{k} - \mu_q\right)}{2M\sigma_q^2},\tag{4.15}$$

where j is the excess degree of a particular node, and $\overline{k}$ is the average remaining degree of the node's neighbors; M is the number of edges in the network; μ_q and σ_q are the mean and standard deviation of the remaining degree distribution of the network, respectively.

4.2.6 Modularity

Modularity measures the *strength* and *quality* of partition of a network into communities or clusters [14,22]. The modularity is the difference between the fraction of edges falling within a community and the expected fraction of such edges in an equivalent network, with edges placed at random. Therefore we can define the modularity measure for partitioning a network into c communities as follows:

$$Q = \sum_{i=1}^{c}(e_{ii} - a_i^2),\tag{4.16}$$

where e_{ii} is the fraction of edges in module i, and a_i is the fraction of edges that have at least one vertice in community i. Mathematically, a_i is defined as follows:

$$a_i = \frac{k_i}{2m} = \sum_j e_{ij},\tag{4.17}$$

where k_i is the degree of the node, and $m = \frac{1}{2}\sum_i k_i$ is the total number of links in the network. On the other hand, e_{ij} is the fraction of edges with one end vertex in community i and the other in community j, and can be represented as

$$e_{ij} = \sum_{uv} \frac{A_{uv}}{2m},\tag{4.18}$$

where A_{uv} represent the adjacency matrix for all $u \in c_i$ and $v \in c_j$. $A_{uv} = 0$ means that there is no link, and $A_{uv} = 1$ means that there is an edge between the two nodes u and v.

The value of Q is bounded between -1 and 1. In general, a large value of Q corresponds to better results.

4.3 Network models

Almost all large-scale networks share some common topological properties, such as *scale-free distributions*, *low average path length*, and *strong community structure*. Based on these properties, real-world networks are classified into the following models.

4.3.1 Random networks

There are two models of a random network. Accordingly, the following are two definitions of random networks commonly used:

1. Erdős-Rényi (ER) Model [9]: The ER model represents an abstract representation of a random network in which a specified probability describes the existence of an edge between each couple of nodes. Formally, a *random graph*, $\mathcal{G}(V, P)$ is a graph with V nodes, where each possible edge has probability P of existence. Consequently, the number of edges in such a graph is a random variable.

2. Gilbert Model [13]: In this model, each pair of V nodes is connected with probability P. Each edge is introduced to a network of nodes V with equal probability $P^{\mathcal{E}}(1-P)^{\binom{|V|}{2}-\mathcal{E}}$, where $\mathcal{E}$ denotes the number of edges in the network.

In a network, any two nodes are likely to show a high clustering coefficient if they have a common third node. However, in the case of Erdős–Rényi [9] and Gilbert [13] random model, the probability of the presence of an edge between any two nodes is independent. Therefore Erdős–Rényi random network has a low clustering coefficient. The average path length of a random network grows logarithmically with the size of the network, i.e., $\langle L \rangle \sim ln(|V|)$. In a random network [13], the probability of a node that has k links follows the binomial distribution and can be defined as

$$P_k = \binom{|V|-1}{k} P^k (1-P)^{|V|-1-k}, \tag{4.19}$$

where V is the set of nodes connected with probability P of a random network $\mathcal{G} = (V, P)$. $\binom{|V|-1}{k}$ represents how many different ways we could pick k links from $V-1$ potential links a node can have. P_k and $(1-P)^{V-1-k}$ represent the probability of links present (k) and missing ($V-1-k$). The shape of this distribution depends on network size (V) and the probability (P). The degree distribution of the random graph follows the Poisson distribution. The average degree ($\langle k \rangle$) specifies the connectivity within random networks, as shown in Fig. 4.5. When $\langle k \rangle < 1$, there exists a large number of small subgraphs. If $\langle k \rangle \gg 1$, it indicates the presence of both large and small subgraphs in the network. The transition phase occurs when $\langle k \rangle = 1$.

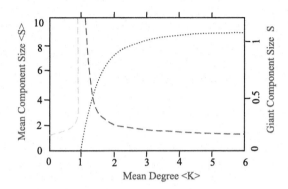

Figure 4.5. Behavior of random network with average degree $\langle k \rangle$.

For a random graph, the average degree z of a vertex is $z = \frac{|\mathcal{V}|(|\mathcal{V}|-1)}{|\mathcal{V}|} \approx np$ for large $\mathcal{V}$. So, once one knows $\mathcal{V}$, any property can be expressed both in terms of P or z. Consequently, this model presents the advantage of summarizing their topological properties in terms of two parameters, $\mathcal{V}$ and p. Briefly, it is possible to recall that for large values of $\mathcal{V}$ (or, alternatively, when $z = 1$), random graphs exhibit a transition phase causing the formation of a so-called *giant component*. A component is a subset of nodes, which are all reachable from other nodes. A giant component, consequently, is the largest component.

The formation of a giant component is a characteristic of many real networks, including biological and social. Despite this, random graphs do not capture the property of high clustering coefficient of real networks. This drawback also appears in metabolic networks as reported in [11]. In that work, authors analyze a metabolic network of *E. coli* by building a graph of interactions, in which vertices represent substrates and products, and edges represent interactions. The clustering coefficient of the network is 0.59, whereas a random graph with the same number of node presents a value of 0.09.

4.3.2 Small-world networks

The small-world phenomenon is indeed very common in many real-world networks. It states that any two individuals in the world can reach out from one individual to another through a path of six or fewer acquaintances between them. This small-world aspect is known as *"small-world effect"*, or *"six degree of separation"*, introduced by the social psychologist Stanley Milgram [32]. The small-world patterns are common in many real-world networks. Small-world networking has many diverse application areas, including brain network [31,35], computing [15,29], earth science [17,34]. The high clustering structural feature of *small-world network* facilitates information or resource sharing among interconnected nodes. Low path length reduces the overall cost during resource sharing throughout the network. It means, spreading of information across the network will be fast in most of the real-world networks. For small-world network, the average path length is proportional to $\log(|\mathcal{V}|)$.

Both regular lattice (a regular graph, where each edge has the same weight, or represents the same distance in Euclidean space) and random networks are unable to show the structural resemblance with most of the real-world networks. The regular lattice has a high clustering coefficient, but it does not follow the small-world effect. Similarly, the random network follows a small-world effect, but it has a low clustering coefficient. Watts and Strogatz [33] introduced a model (known as the WS model), which describes the transformation of a regular network to random networks. According to Watts and Strogatz, *small-world* networks are the intermediate networks present between the regular and random networks, which have *high clustering coefficient* like the regular lattice and *low path length* like random networks (*see* Fig. 4.6). The WS model can be built on a one-dimensional regular lattice by *rewiring* the connection among the nodes. In this *rewiring* procedure, the one end of a connection is assigned to another new location (node), chosen randomly from the network. Furthermore, it should follow some constraints, such as there should not be any double connection between any two nodes or presence of self-edge. Pictorially the fact is illustrated in Fig. 4.6. During the *rewiring* process, for a small change (or no change) of *rewiring* probability, there is an insignificant change (or no change) in the clustering coefficient value from its initial value ($C(P) \sim C(0)$). Whereas, there is an expeditious decrease in average path length and is in the same order as the one for the random network ($L(P) \ll L(0)$).

Newman and Watts [23] proposed a variant model of *small-world*, known as NW small-world model. In this model, during *rewiring* process, the connection present among the nodes are not broken down. Instead, a new edge with probability P is added between any pair of nodes. Like the WS model, it also follows the same constraints. Nodes are not allowed to be coupled to other nodes more than once, or to couple itself. For an adequately small value of P and for a large value of V, the behavior of both WS–NW model is substantially identical. The main advantage of NW model over the WS model is that it is comparatively easier to analyze, because it does not form isolated subgraphs.

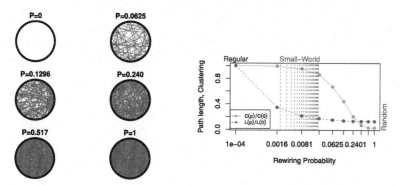

Figure 4.6. Watts and Strogatz small-world model. A regular ring lattice of 200 nodes is rewired with varying probability (P) ranging from 0 to 1. For $P = 0$, we have a regular lattice with high clustering coefficient and low path length. For $P = 1$, we have low clustering coefficient as well as low path length, same as those for random networks. Between $P = 0$ and $P = 1$, the region where we obtained high clustering coefficient and simultaneously low path length are small-world networks.

4.3.3 Scale-free networks

In scale-free networks, almost all nodes have a roughly equal number of connections. The degree distribution for such networks follow Poisson distribution, with a peak at an average value. A recent study on large complex networks, such as the Internet, social networks, WWW, and metabolic networks, highlighted the inhomogeneous nature of the real-world networks. Most of the real-world networks usually follow the properties of scale-free network [3], where a large number of nodes have a low degree, and, contrarily, there exist few high-degree (hub) nodes. This anomaly in real-world networks is being introduced by Barabási and Albert (BA) [3]. Barabási network model deviates from the conventional network model, stating that traditional model only considers the alteration of connections. Addition of new nodes and deletion of existing nodes is not possible throughout the network creation process. However, most of the real-world networks are evolving in nature, where nodes are continuously added or deleted over time. Furthermore, Barabási and Albert also pointed out the consideration of uniform probabilities adopted by network models, such as NW and WS model during the creation of new connections.

In real-world scenarios, this perception becomes unrealistic. For example, a highly cited research article is likely to get more and more citation in the future in comparison to a low-cited research paper. Similarly, the fan circle of well-known (popular) people, such as actors, sportsmen, increases over time than normal people. This phenomenon of *preferential attachment bias* is known as *"rich-get-richer"*, which is overlooked by other models. Due to the preferential attachment property, low degree nodes tend to connect more with the hub or core nodes and connectivity becomes sparse as it moves towards the boundary.

Barabási and Albert model starts with a few number of nodes (m_0), and after every interval of time t, a new node is introduced to the network. These newly added nodes are connected to/from $m \leq m_0$ existing nodes. The probability of preferential attachment Π_j that a new node v_i to get connected with an existing node v_j, randomly chosen over m existing nodes, is measured by the degree of v_j (i.e., k_j) such that $\Pi_j = \frac{k_j}{\sum_{p=1}^{|\mathcal{V}|} k_p}$, where $|\mathcal{V}|$ denotes the total number of nodes.

For the real-world networks, another intriguing aspect of preferential attachment $\Pi(k)$ is that it has a nonzero value towards an isolated node, i.e., $\Pi(0) \neq 0$. Therefore the preferential attachment $\Pi(k)$ for the real-world networks can be generalized as $\Pi(k) = \mathcal{A} + k^\alpha$, where $\mathcal{A}$ denotes the initial pulling competence of a node and α is the power law exponent. Since, scale-free networks are a kind of ultra-small world [6,7] networks, therefore average path length of scale-free networks are proportional to $\log(\log(\mathcal{V}))$. The clustering coefficient of scale-free networks follows the *power law*, which is inversely proportional to the node degree.

Fig. 4.7, represents the basic properties exhibited by random, small-world and scale-free networks. In this experiment, we illustrate how network size (N) affects the average path length (left) and the clustering coefficient (middle) in ER/WS/BA network model. Fig. 4.7 (right) shows that the scale-free network follows the power-law distribution, whereas the small-world network follows the Poisson distribution (peak), and the random network shows a bell curve distribution.

4.3.4 Other models

In addition to the above three models, which are considered major graph models, few more models have been introduced recently.

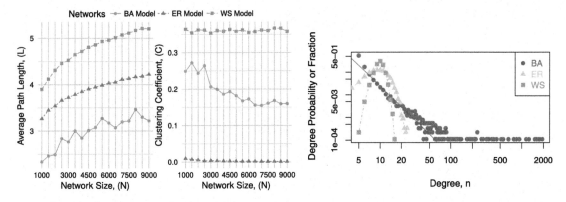

Figure 4.7. Depicts the behavior of network model ER/WS/BA with respect to *average path length* (left), *clustering coefficient* (middle), and *degree distribution* (right) as a function of network size, N. For the Erdős–Rényi network, $k = 10$, and thus $p = 10/N$. For the Watts–Strogatz undirected network, $P = 0.1$, and the mean degree is $k = 5$. For the Barabási–Albert undirected network, $m = m_o = 5$.

Geometric random graph model

A geometric random graph (GRR) $\mathcal{G}(\mathcal{V}, r)$ is a graph [25] whose nodes are points in a metric space, which are connected by an edge if their distance is below a threshold value r called radius. Formally, let $u, v \in \mathcal{V}$, the edge set is $\mathcal{E} = \{\{u, v\} | (u, v \in \mathcal{V}) \wedge (0 < \|u - v\| < r)\}$, where $\|.\|$ is a defined distance norm. Generally, a two-dimensional space is considered, and norms are the well known Manhattan or Euclidean distance, and the radius takes values in $(0, 1)$.

Thus a random geometric graph $\mathcal{G}(n, r)$ is a generalization of this model, in which nodes correspond to n points in a metric space. Clearly, these points are distributed uniformly and independently. Properties of these graphs have been studied when $n \to \infty$. Surprisingly, certain properties of these graphs appear when a specific number of nodes is reached.

Sticky model

The stickiness index model was introduced in [28] and it is related to the number and distribution of distinct binding domains (i.e., the regions of the structure of the proteins that are responsible for the interactions). The model, developed starting from some complex previous ones, is based on the assumption that the abundance and popularity of binding domains on a protein may be summarized in a single index related to its normalized degree, named the stickiness index. The models are based on two considerations: (i) a high degree of a protein is related to the presence of many binding domains; (ii) a pair of proteins is more likely to interact if both have high stickiness indices. Therefore for each pair of proteins, the product of the two stickiness defines the probability of interaction.

Hypergeometric graph model

The hypergeometric graph model [18] is based on the development of a geometric framework to study the structure and function of complex networks. The main assumption of this model is that hyperbolic geometry underlies complex biological networks. As a consequence, degree distributions and strong clustering are motivated by the negative curvature and metric property of the underlying hyperbolic geometry.

4.3.5 Characteristic analysis of few real networks

We experimented with few real-life networks and report their average degree, average path length, maximum distance (diameter) and clustering coefficient in Table 4.1.

Table 4.1 Characteristics of various real-world networks. Each network has N number of nodes, E number of edges, average degree $\langle K \rangle$, average path length L, maximum distance (diameter) L_{max}, and clustering coefficient, CC.

Network	$\mathcal{N}$	$\mathcal{E}$	$\langle K \rangle$	L	L_{max}	C
Zachary karate club	34	78	4.5	2.44	5	0.25
Dolphins	62	159	5.12	3.45	8	0.309
E. Coli Metabolism	1,039	5,802	5.58	2.98	8	0.59
Protein Interactions	2,018	2,930	2.9	5.61	14	0.023
Power Grid	4,941	6,594	2.67	18.99	46	0.107
Science Collaboration	23,133	93,439	8.08	5.35	15	0.32
arXiv hep-ph	28,093	3,148,447	327.26	2.83	9	0.280
Mobile Phone Calls	36,595	91,826	2.51	11.72	39	0.15
Email	57,194	103,731	1.81	5.88	18	0.032
Facebook friendships	63,731	817,035	25.64	4.31	15	0.148
Internet	192,244	609,066	6.34	6.98	26	0.24
DBLP co-authorship	317,080	1,049,866	6.62	6.75	23	0.306
WWW	325,729	1,497,134	4.6	11.27	93	0.11
Amazon (MDS)	334,863	925,872	5.52	11.73	47	0.205
Citation Network	449,673	4,707,958	10.43	11.21	42	0.43
Actor Network	702,388	29,397,908	83.71	3.91	14	0.79
LiveJournal	4,847,571	68,475,391	28.25	5.48	20	0.118

4.4 Graph modeling in R

Given any arbitrary graph, it is easy to process and analyze it using igraph package in R. Here, we first demonstrate how to create complex graph models synthetically. Next, we analyze their topological properties.

4.4.1 Modeling complex network

Ready-to-use R functions are available to simulate various complex network models. The following sample codes will show how to use them for generating such networks synthetically before starting any analysis (see Fig. 4.8, Fig. 4.9, and Fig. 4.10):

```
Random network (Erdos–Renyi)

net=sample_gnm(n=200, m=60) \\n, m -> represent number of
    nodes and edges respectively

plot(net, vertex.size=6, vertex.label=NA)
```

or

```
Random network (Erdos–Renyi)

z=50 \\ size of the network
net=erdos.renyi.game(z, 0.1)
plot(net, vertex.size=6, vertex.label=NA)
```

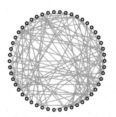

Figure 4.8. Random network (Erdos–Renyi).

Small-world network (Watts–Strogatz)

```
net=sample_smallworld(dim=2, size=7, ngh=2,p=0.2)  \\ dim
    -> dimensions; p -> probability; ngh ->
    neighborhood

plot(net, vertex.size=6, vertex.label=NA, layout=
    layout_in_circle)
```

or

```
z=60
net=rewire.edges(erdos.renyi.game(z, 0.1), prob=0.8)
plot(net, vertex.size=6, vertex.label=NA)
```

Figure 4.9. Small-world network (Watts–Strogatz).

Scale-free network (Barabasi–Albert)

```
net=sample_pa(n=200, power=1, m=1, directed=F) \\ n,m ->
    number of nodes and edges; power -> power of
    attachment

plot(net, vertex.size=10, vertex.label=NA)
```

or

```
z=40
net=barabasi.game(z)
plot(net, vertex.size=6, vertex.label=NA)
```

Figure 4.10. Scale-free network (Barabasi–Albert).

4.4.2 Topological analysis

Following are the sample R codes and their outputs for topological analysis (see Fig. 4.11, Fig. 4.12, Fig. 4.13, and Fig. 4.14):

Degree distribution

```
z=150
net=erdos.renyi.game(z, 0.1)   \\ 0.1 -> average degree

deg=degree(net, mode="all")
dlist=degree.distribution(net, mode="all", cumulative=T)
plot(x=0:max(deg), y=dlist, pch=16, cex=1.2, col=c(1:20),
xlab="Degree", ylab="Cumulative Frequency")
lines(x=0:max(deg), y=dlist, col="blue")
```

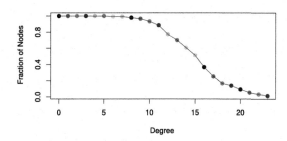

Figure 4.11. Degree distribution.

Average Path length

```
links=read.csv("igraph.csv", header=T, as.is=T) \\ Read
    your network file

net=graph.data.frame(, directed=F)   \\ network
    representation

Avg=average.path.length(net, unconnected=T)
print(Avg)
```

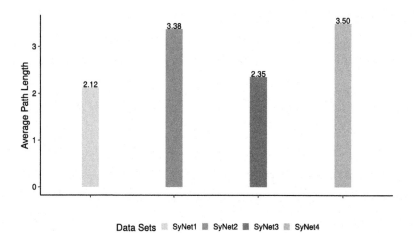

Figure 4.12. Average path length.

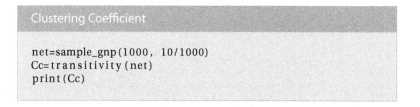

```
net=sample_gnp(1000, 10/1000)
Cc=transitivity(net)
print(Cc)
```

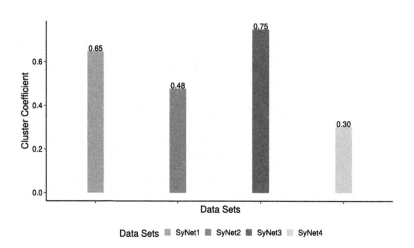

Figure 4.13. Plotting of clustering coefficients.

```
install.packages("brainGraph")
rich.club.coeff(g, k=1, weighted=FALSE) \\ g -> The graph
    of interest. k -> The minimum degree for including
    a vertex (default: 1)

rich.club.norm(g, N=100, rand=NULL, ...) \\ N -> The
    number of random graphs to generate (default: 100),
    rand -> A list of igraph graph objects, if random
    graphs have already been generated (default: NULL),
    ... -> Other parameters (passed to rich.club.coeff)
```

Assortativity

```
assortativity(graph, types1, types2=NULL, directed=TRUE)
assortativity_nominal(graph, types, directed=TRUE)
assortativity_degree(graph, directed=TRUE)
```

Modularity

```
modularity(graph, membership, weights=NULL)
\\ graph -> The input graph. membership -> Numeric vector
    , for each vertex it gives its community. The
    communities are numbered from zero. weights -> If
    not NULL then a numeric vector giving edge weights.
```

Examples

```
g=graph.full(5) %du% graph.full(5) %du% graph.full(5)
g=add.edges(g, c(0,5, 0,10, 5, 10))
wtc=walktrap.community(g)
memb=community.to.membership(g, wtc$merges, steps=12)
modularity(g, memb$membership)
```

Figure 4.14. R codes for computing Rich club, assortativity and modularity.

4.5 Summary

We discussed here different network topological properties possessed by real-world networks. Based on their properties, they may be classified into different network models. Particularly, all real-world networks can either be classified as scale-free networks or small-world networks. In recent years, the examination of the structural features of real-world networks has become a new focal research area among computer scientists. Neither node-wise analysis nor the analysis of the entire network at once is suitable for the large networks. Hence, subgrouping of similar kinds offers an intuitive solution for the large network analysis, and it presents a persuasive way to visualize networks at a different resolution. During the last few years, researchers have put lots of emphasis on identifying subgroups (or communities) in real-world networks. Finding overlapping, intrinsic communities is a current need. The problem becomes more challenging when considering the dynamic nature of the graph, as real networks are continuously growing and shrinking too.

Acknowledgment

This chapter is coauthored by Dr. Keshab Nath. He worked in the area of social network analysis as a part of doctoral research.

References

1. Réka Albert, Albert-László Barabási, Topology of evolving networks: local events and universality, Physical Review Letters 85 (24) (2000) 5234.
2. Réka Albert, Albert-László Barabási, Statistical mechanics of complex networks, Reviews of Modern Physics 74 (1) (2002) 47.
3. Albert-László Barabási, Réka Albert, Emergence of scaling in random networks, Science 286 (5439) (1999) 509–512.
4. Albert-László Barabási, et al., Network Science, Cambridge University Press, 2016.
5. Gian Paolo Clemente, Rosanna Grassi, Directed clustering in weighted networks: a new perspective, Chaos, Solitons and Fractals 107 (2018) 26–38.
6. Reuven Cohen, Shlomo Havlin, Scale-free networks are ultrasmall, Physical Review Letters 90 (5) (2003) 058701.
7. Reuven Cohen, Shlomo Havlin, Daniel Ben-Avraham, Structural properties of scale-free networks, in: Handbook of Graphs and Networks, 2002.
8. Vittoria Colizza, Alessandro Flammini, M. Angeles Serrano, Alessandro Vespignani, Detecting rich-club ordering in complex networks, Nature Physics 2 (2) (2006) 110.
9. P. Erdős, A. Rényi, On random graphs I, Publicationes Mathematicae Debrecen 6 (1959) 290–297.
10. Giorgio Fagiolo, Clustering in complex directed networks, Physical Review E 76 (2) (2007) 026107.
11. David A. Fell, Andreas Wagner, The small world of metabolism, Nature Biotechnology 18 (11) (2000) 1121.
12. Jacob G. Foster, David V. Foster, Peter Grassberger, Maya Paczuski, Edge direction and the structure of networks, Proceedings of the National Academy of Sciences 107 (24) (2010) 10815–10820.
13. E.N. Gilbert, Random graphs, The Annals of Mathematical Statistics 30 (4) (1959) 1141–1144.

14. Michelle Girvan, Mark EJ Newman, Community structure in social and biological networks, Proceedings of the National Academy of Sciences 99 (12) (2002) 7821–7826.
15. Robert Hillard, Information-Driven Business: How to Manage Data and Information for Maximum Advantage, John Wiley & Sons, 2010.
16. Paul W. Holland, Samuel Leinhardt, Transitivity in structural models of small groups, Comparative Group Studies 2 (2) (1971) 107–124.
17. A. Jiménez, K.F. Tiampo, A.M. Posadas, Small world in a seismic network: the California case, Nonlinear Processes in Geophysics 15 (3) (2008) 389–395.
18. Dmitri Krioukov, Fragkiskos Papadopoulos, Maksim Kitsak, Amin Vahdat, Marián Boguná, Hyperbolic geometry of complex networks, Physical Review E 82 (3) (2010) 036106.
19. Matthieu Latapy, Clémence Magnien, Nathalie Del Vecchio, Basic notions for the analysis of large two-mode networks, Social Networks 30 (1) (2008) 31–48.
20. R. Duncan Luce, Albert D. Perry, A method of matrix analysis of group structure, Psychometrika 14 (2) (1949) 95–116.
21. Mark E.J. Newman, Assortative mixing in networks, Physical Review Letters 89 (20) (2002) 208701.
22. Mark E.J. Newman, Michelle Girvan, Finding and evaluating community structure in networks, Physical Review E 69 (2) (2004) 026113.
23. Mark E.J. Newman, Duncan J. Watts, Renormalization group analysis of the small-world network model, Physics Letters A 263 (4–6) (1999) 341–346.
24. Tore Opsahl, Triadic closure in two-mode networks: redefining the global and local clustering coefficients, Social Networks 35 (2) (2013) 159–167.
25. Mathew Penrose, et al., Random Geometric Graphs, vol. 5, Oxford University Press, 2003.
26. Mahendra Piraveenan, Mikhail Prokopenko, Albert Zomaya, Assortative mixing in directed biological networks, IEEE/ACM Transactions on Computational Biology and Bioinformatics 9 (1) (2012) 66–78.
27. Mahendra Piraveenan, Mikhail Prokopenko, A.Y. Zomaya, Local assortativeness in scale-free networks, Europhysics Letters 84 (2) (2008) 28002.
28. Nataša Pržulj, Desmond J. Higham, Modelling protein–protein interaction networks via a stickiness index, Journal of the Royal Society Interface 3 (10) (2006) 711–716.
29. Oskar Sandberg, Searching in a small world, PhD thesis, Chalmers tekniska högskola, 2005.
30. Sara Nadiv Soffer, Alexei Vazquez, Network clustering coefficient without degree-correlation biases, Physical Review E 71 (5) (2005) 057101.
31. Olaf Sporns, Dante R. Chialvo, Marcus Kaiser, Claus C. Hilgetag, Organization, development and function of complex brain networks, Trends in Cognitive Sciences 8 (9) (2004) 418–425.
32. Jeffrey Travers, Stanley Milgram, The small world problem, Phychology Today 1 (1) (1967) 61–67.
33. Duncan J. Watts, Steven H. Strogatz, Collective dynamics of small-world networks, Nature 393 (6684) (1998) 440.
34. Xin-She Yang, Small-world networks in geophysics, Geophysical Research Letters 28 (13) (2001) 2549–2552.
35. Shan Yu, Debin Huang, Wolf Singer, Danko Nikolić, A small world of neuronal synchrony, Cerebral Cortex 18 (12) (2008) 2891–2901.
36. Shi Zhou, Raúl J. Mondragón, The rich-club phenomenon in the Internet topology, IEEE Communications Letters 8 (3) (2004) 180–182.

5

Biological network databases

Contents

Data about biological networks are the essential ingredient for data mining tasks and comparison of systems. Therefore the development of adequate storage and querying systems is a crucial challenge for bioinformatics.

The first storage systems were implemented using a relational database and some times on flat files. Despite the simplicity, both forms present some drawbacks related to access times, as well

Biological Network Analysis. https://doi.org/10.1016/B978-0-12-819350-1.00011-6

as querying capabilities. Therefore in recent years, many NoSQL databases were introduced and used. This chapter introduces the main concepts of graph databases. Then some experiences are discussed.

5.1 No-SQL and graph databases

Relational databases (RDB) were developed in the early 1970s, and they rapidly became the standard for database management systems. RDBs are currently the best choice for modeling data with relational properties, whereas more recently the production of data with low structures and big dimensions (e.g., biological network data, social network data, and, in general, big data) is growing.

Consequently, the use of traditional RDB systems has some drawbacks, i.e., to obtain complex information from multiple relations, RDB sometimes needs to perform expensive SQL (Structured Query Language) join operation to merge two or more relations at the same time. To mitigate, besides traditional data storage format, other data storage formats have been proposed, often referred to as No-SQL (not only SQL) databases. There exist many different structures of No-SQL databases, such as key-value pairs, document-oriented, time series, and we focus in particular on graph databases [3].

Among the others, we focus here on graph databases (GdB), i.e., a database that uses a graph structure for expressing queries based on nodes, edges and properties for storing attributes related to nodes and edges. The core of a graph database model is the concept of graph used to associate data items stored as nodes using tips representing the relationship among them. Relationships link data together in an easy way, and it results faster data retrieval (i.e., with constant time in many cases).

In a GdB, nodes represent entities, such as proteins, biological molecules, people or patients. Each node may be seen as the translation of a row (or record) of a relational database. Similarly, edges connecting nodes represent relationships among two records, and they can either be directed or undirected. When graphs are directed, the direction of the edge represents, in general, a different meaning. In a GdB, edges constitute the key concept, since they represent an abstraction that is not representable easily in the relational model. Each node may have a set of associated properties, i.e., the GdB represents a protein interaction network; each node may be associated with the name of the protein, cross-referenced to an external database and other biological information.

GdB uses two main conceptual models: the graph property model and the resource description framework (RDF) model.

In the first model, the GdB is based on the use of a set of nodes, relationships, properties, and labels. Each node is associated with a label, and its properties are stored using key/value pairs. Each edge is, in general, directed and may have associated properties represented as key/value pairs. Direct storage of relationships allows a constant-time traversal. Fig. 5.1 depicts this model.

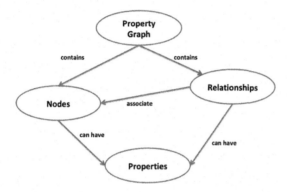

Figure 5.1. A graph property data model is composed by a set of nodes and edges (relationships). Each node is associated with a label, and its properties are stored using key/value pairs. Each edge is, in general, directed and may have associated properties represented as key/value pairs.

In a resource description framework graph model, each information is represented with a separate node in a subject-predicate-object model, known as triples. For instance, one way to represent the concept of the protein X has the function Y as triplet: X (the subject), has a function (the predicate), Y (the object). Diversely in the graph model, X is a node and function; Y is an associated key/value pair as depicted in Fig. 5.2. It should be noted that each additional information is represented by a distinct node. A node may be left blank, a literal and/or be identified by a URIref. An edge may also be identified by a URIref.

Graph databases are usually queried using **Cypher** [8], an ad hoc developed query language. Cypher is a query language for graph databases based on the property of graph model. It was designed first only for Neo4j database,[1] and successively it has become a de facto standard for several graph databases, both academic and commercial ones. Cypher is actually at version 9 maintained by the openCypher Implementers Group that is also in charge of developing version 10.

[1]http://neo4j.org.

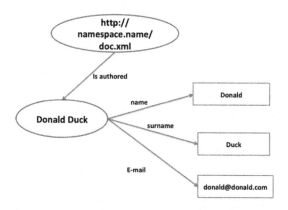

Figure 5.2. An RDF data mode that represents that Donald Duck is author of the document identified by the URI http://namespace.name/doc.xml. Donald Duck also has three properties: name, surname, and e-mail represented by rectangles, and linked to Donald Duck by three distinct edges.

Cypher is a declarative language, and it provides commands for querying and modifying data as well as for specifying schemas. Cypher queries take as input a property graph and return a table representing the output. A query is built as a linear composition of many clauses.

The central concept in Cypher queries is a pattern matching expressed by the clause **MATCH**. An example of a pattern expressed in Cypher is $(a) - [r] \rightarrow (b)$, where a and b are two nodes and $[r]$ is an edge connecting them. A cypher query is composed of many clauses that are merged together linearly, helping the user to form a query quickly. The output is a set of tables that can both expand the number of fields and add new tuples.

Example of accessible graph databases, as reported in database Ranking, provided by the website db-engines.com, are Neo4j, Microsoft Azure Cosmos DB, OrientDB, ArangoDB, and Virtuoso.

Neo4JDatabase   Neo4j[2] is ranked as the most popular graph database. Neo4j analyzes and traverses data in real-time and gives the results very fast. Neo4j is an open-source NoSQL graph database written in Java and Scala. Neo4j provides ACID transaction compliance, cluster support, runtime failover, high availability, and high speed querying through traversals. It may scale rapidly to billions of nodes and edges. It has a big community that can help. Neo4j is based on the property graph model, i.e., the stored graph has its nodes connected through relationships. Both nodes and links store data in key-value pairs known as properties. Properties

[2]http://neo4j.org.

can be added or removed when necessary. Querying the data is based on the Gremlin traversal language and Cypher Query Language.

Microsoft Azure Cosmos DB: Azure Cosmos DB[3] is Microsoft proprietary distributed, multimodel database service. It was launched in 2017, and it is classified as a NoSQL database, since it may represent data using different approaches. In particular, it may represent data through Gremlin language [24]. Gremlin is a graph traversal language and developed by Apache TinkerPop of the Apache Software Foundation.

OrientDB database: OrientDB[4] is an open-source NoSQL database management system developed in Java by Orient Technologies LTD. OrientDB has a hybrid nature, since it collects features of both document and graph databases. In the so-called graph mode, objects are linked together through referenced relationships that join a start vertex and an end vertex. Properties may be associated with both relationships and vertex. A relational model may be stored into OrientDB by representing it through a document-graph model. OrientDB supports an extended version of SQL, enabling CRUD and query operations. It is distributed under the open source Apache2 license.

ArangoDB ArangoDB[5] is a multimodel No-SQL database that unifies graph, document, and key/value models through a single query language.

5.1.1 No-SQL databases in bioinformatics

Bio4J: Bio4J[6] is a graph database that was developed for the integration of semantically rich biological data using typed graph models. Authors developed a novel graph data-model [30] and then used it to integrate a large number of publicly available data related to proteins by using a set of linked graphs.

The BioGraph database: BiographDB [23] is a graph database that integrates heterogeneous biological resources available through a web application named Biograph [22]. The BiographDB integrates many sources as depicted in the following table (data are related to the data of the publication of this book):

---

[3]https://azure.microsoft.com/it-it/services/cosmos-db/.
[4]https://orientdb.com.
[5]www.arangodb.com.
[6]www.bio4j.com.

Data source	Version/Date
NCBI Entrez Gene	09/05/2017
UniProt Swiss-Prot	2017 04
HGNC	10/05/2017
Gene Ontology	09/05/2017
Reactome	v59
miRBase	Release 21
miRCancer	December 2016
microRNA.org	August 2010
miRTarBase	6.1
miRNASNP	2.0

5.1.2 Pros and cons of using No-SQL databases

The pros and cons of using a graph database instead of relational databases is thus an important research area. Have and Jenses [11] compared the use of Neo4J databases concerning PostgreSQL on the human interaction network imported from the STRING database. The network used in the experiment has 20,140 proteins and 2.2 million interactions.

Neo4J stores the edges as pointers between two nodes, thus enabling the traversal of nodes in constant time. Properties associated with nodes and edges (such as node name, confidence scores of interactions, source of communications, etc.) are stored together with nodes and edges, since Neo4J uses the property graph model. In such a model, data is organized as nodes, relationships, and properties (data stored on the nodes or relationships). Authors [11] stored the graph in PostgreSQL[7] as a table of node pairs. The table is indexed, therefore it can be traversed in logarithmic or constant time based on the index used.

The comparison of databases has been made measuring the speed of Cypher and SQL queries for solving three problems:

- finding immediate neighbors and their interactions,
- finding the best scoring path between two proteins,
- finding the shortest path between them.

Authors measured a great speedup of No-SQL over a relational database. Despite this, it does not necessarily imply that the nonrelational databases are the best choice always. They note that when queries are formulated in terms of paths, then

[7] www.postgresql.org.

graph databases are more concise and clear. Conversely, relational databases are more evident when set operations are needed.

A plethora of databases is available publicly and privately, storing extensive biological experimental data maintained in various database formats. With the advent of high throughput experimental setup and advanced database technologies, it is now possible to generate, store, and access a high volume of experimental data in various repositories conveniently. Practical data analysis is now possible to elucidate previously unknown biological facts on applying various data analytic and inference tools on the stored data.

Next, we discuss few popularly used data sources for three biological networks: gene interactions, protein interactions, and brain connectomes.

5.2 Genetic interaction network databases

The number of reported genetic interactions are relatively less in comparison to other biological networks. This may due to the involvement of various indirect factors that determine true physical interactions, and hence not possible to elucidate true relationship based on a single source of information. Majority of the interactions are predicted *in silico* and reported in the databases. Microarray and RNA sequence reads are the most popularly used data sources for predicting such interactions. However, they are sensitive towards the quality, reliability, and availability of the data. Also, the interactions largely depend on the merit of the inference method used. Below, we discuss a few databases dealing with gene-gene relationship networks.

5.2.1 RegNetwork (regulatory network repository)

This repository [21] focuses on five varied types of human and mouse transcriptional and posttranscriptional regulatory relation. It consists of detailed information about various combination of synergic organizational relationships among TFs, miRNAs and genes.

5.2.2 TRRUST v2

Transcriptional regulatory relationships unraveled by sentence based text mining [10], is a database which consists of human and mouse transcriptional regulatory networks. It comprises of 8,444 and 6,552 TF-target regulatory relationships of 800 human TFs and 828 mouse TFs, respectively.

5.2.3 TRED

Transcriptional regulatory element database [29], consists of a number of promoters and genes of human, mouse, and rat. This database focuses on GRN's for each TF-target gene pairs involved in cancer. This database also consists of other features: it contains the genome-wide promoter annotation, gene transcriptional regulation. It also provides an interface, which is user-friendly for extraction of data for all the three species.

5.2.4 BioGRID

Biological general repository [26] for interaction datasets consists of interaction of 70 different organisms, such as the horse, tomato, and castor bean. This repository searches 71,178 for 1,753,686 protein and genetic interactions, 28,093 chemical associations, and 874,796 posttranslational modifications from major model organism species.

5.2.5 miRTarBase

It is a database [14], which curates validated microRNA-target interactions. This database is a house for 657 miRNAs and 2297 target genes among 17 species, such as human, mouse, chicken, sheep, etc. This database is a house for 4,22517 curated MTIs (miRNA-target interactions) from 4076 miRNAs and 23 054 target genes collected from over 8500 articles. The database also observes the functionality of target genes playing a role in human MTIs using gene ontology and KEGG pathway enrichment annotation using the DAVID.[8]

5.2.6 KEGG pathway database

Kyoto encyclopedia of genes and genomes [16,17] is developed by Kanehisa Laboratories. This database consists of different types of human cancer disease dataset. This database also comprises of pathway maps for the molecular systems in both standard and perturbed states. This database provides molecular-level information of cells.

Table 5.1 consists of some of the details about the respective databases.

---

[8]https://david.ncifcrf.gov/.

Table 5.1 Few publicly available genomic interaction databases.

Database	No. of species in the database	Last updated	Download link
RegNetwork	2	2019/04/02	http://www.regnetworkweb.org/home.jsp
TRRUST v2	2	2018/04/16	https://www.grnpedia.org/trrust/
TRED	3	2007/02/27	https://www.hsls.pitt.edu/obrc/index.php?page=URL1103823836
BioGRID	70	2019/12/01	https://thebiogrid.org/
miRTarBase	17	2019/09/15	http://mirtarbase.mbc.nctu.edu.tw/php/index.php
KEGG	1	2019/10/01	https://www.genome.jp/kegg/

5.3 Protein-protein network databases

The management of protein-protein interaction (PPI) data presents similar issues as those faced in other domains, i.e., PPI data need to be stored, exchanged, queried, and analyzed. PPI data are the constitutive building blocks for protein interaction networks (PINs). This section discusses main phases and issues of PPI data management [6].

Regarding PPI data storage, main efforts were devoted to the definition of standards for data exchange, such as HUPO PSI-MI, but currently, PPI data are stored as large sets of binary interactions, without taking into account XML-based languages and related XML databases. The storage of PPI data could exploit some already developed storage systems for other graph-based data, such as the triple stores used for storing RDF data or the emerging graph databases [4]. In graph databases, schema and instances are modeled as graphs, and data manipulation is expressed by graph-oriented operations. A graph database proposal for genomics is reported,[9] and a project for biochemical pathways is reported in [7].

Moreover, a naming mechanism to identify interactions in a unique way has not been yet been developed, and (binary) interactions are named by naming the interacting proteins.

Also, PPI data querying could benefit from semi-structured or graph databases as summarized below; existing PPI data offer only very simple retrieval mechanisms allowing the retrieval of proteins interacting with a target protein. Current PPI databases surveyed in this paper do not offer sophisticated query mechanisms based on graph manipulation, but, on the other hand, they con-

[9]http://www.xweave.com/people/mgraves/pubs/.

stitute the only available structured repository for interaction data and allow an easy sharing and annotation of such data. Moreover, all the existing databases go beyond the storing of the interaction, but integrates it with functional annotations, sequence information and references to corresponding genes. Finally, they generally provide some visualization tools that presents a subset of interactions in a comprehensive graph.

5.3.1 The database of interacting proteins (DIP)

The database of interacting proteins (DIP)[10] [28] contains experimentally verified PPI of different organisms. It is implemented as a relational database. Each entry of DIP stores some general information about proteins (e.g., gene name, cellular localization) and the cross-references to different databases, as well as information about experimental procedures and individual experiments. Finally, each interaction is identified by a unique code. Interactions stored in DIP are mandatorily described in peer-reviewed journals, and the entry-process is manual. Queries can be formulated via a web interface in both interactive and batch modes. In the batch mode, a user can download a subset of DIP choosing different formats (e.g., XML-based or tab-delimited format).

5.3.2 Biomolecular interaction network database (BIND)

The biomolecular interaction network database (BIND)[11] [2] contains protein interactions annotated with molecular function information extracted from literature. It is based on three main types of data records: interaction, molecular complex, and pathway. An interaction record stores a description of the reaction event between two objects. Molecular complexes are stored through the use of interactions, temporally sorted, producing them. When the reactions generating a complex are unknown, the complex is defined more loosely. A pathway, defined as a network of interactions usually mediating some cellular functions, is described as a series of reactions with information, such as cell cycle and associated phenotypes.

The database permits different modes of search: using identifiers from other biological databases, or by using specific fields, such as literature information, molecule structure, and gene information, including functions. The extracted information can be displayed with a BIND interaction viewer. Networks are rendered

---

[10]http://dip.doe-mbi.ucla.edu/.
[11]http://www.bind.ca/.

as graphs, where nodes, representing molecules, are labeled with some ontological information.

5.3.3 IntAct

The IntAct [12][12] database is a database of interactions that is based completely on open-source software. It contains not only protein interactions data, but also DNA and molecular interaction data. IntAct uses a set of controlled vocabularies and ontologies to provide a semantically consistent annotation method. A researcher can submit an interaction, using the PSI-MI format [13], by sending an e-mail to the database curators.

5.3.4 Online predicted human interaction database (OPHID)

The online predicted human interaction database (OPHID)[13] [5] contains predicted interactions between human proteins. It combines PPI derived from literature and from databases, with predictions made from other organisms. OPHID can be queried by using single or multiple protein IDs, and the results can be visualized using its graph visualization program. The database is freely available for academic purposes.

5.3.5 Prediction of interactome database (POINT)

The prediction of interactome database (POINT)[14] [15] stores predicted interactions of human proteins derived from available orthologous interactions datasets. This database uses interactions of *worm, fly,* and *yeast* proteins as a starting point for prediction, then it projects them to the human orthologous. It integrates several publicly accessible databases containing protein-protein interactions of the mouse, fruit fly, worm, and yeast. Human interactions are thus predicted, starting from orthologs proteins.

5.3.6 Integrated network database (IntNetDB)

The integrated network database (IntNetDB)[15] integrates various types of functional data by applying a probabilistic model, and then predicts protein to protein interaction. Users can search interactions by entering a variety of gene identifiers for different

---

[12]http://www.ebi.ac.uk/intact/.
[13]http://ophid.utoronto.ca/ophid/.
[14]http://point.bioinformatics.tw.
[15]http://hanlab.genetics.ac.cn/IntNetDB.htm.

organisms. IntNetDB also provides a tool to extract the highest connected network neighborhoods topologically from a specific network for further exploration and research.

5.3.7 Search tool for the retrieval of interacting genes/proteins (STRING)

The search tool for the retrieval of interacting genes/proteins database (STRING)[16] is a repository that aims to bring together the biochemical associations among protein to protein, protein to DNA, and DNA to DNA. The database can be accessed on the website by specifying a protein identifier, or by inserting the primary sequence of a protein. Free PPI database sources are shown in Table 5.2.

Table 5.2 A summary of few PPI databases.

Sl. No.	Database	No. of species	No. of interactors	No. of interactions	Link
1.	MINT	611	25,530	125,464	http://mint.bio.uniroma2.it/
2.	HPRD	1	30,047	41,327	http://www.hprd.org/
3.	BioGRID	62	65,617	1,423,105	http://thebiogrid.org/
4.	MatrixDB	1	14,533	26,954	http://matrixdb.univ-lyon1.fr/
5.	IntACT	275	98,289	720,711	http://www.ebi.ac.uk/intact/
6.	InnateDB	4	25,110	367,496	http://www.innatedb.com/
7.	DIP	372	95,742	720,711	http://dip.doe-mbi.ucla.edu/dip/Main.cgi
8.	STRING	2031	964,376	1,380,838,440	http://string-db.org/
9.	Mentha	8	88,130	648,080	http://mentha.uniroma2.it/index.php
10.	HAPPI	1	23,060	2,922,202	http://discovery.informatics.uab.edu/HAPPI/
11.	IsoBase	5	87,737	114,897	http://cb.csail.mit.edu/cb/mna/isobase/
12.	APID	8	93912	754895	http://cicblade.dep.usal.es:8080/APID/init.action

5.4 Databases of brain networks

Recently, there has been a growing interest in databases storing only brain networks. These datasets often contain both raw images data as well as brain graphs extracted from images.

[16]http://string.embl.de/.

5.4.1 The healthy brain network (HBN) initiative and its dataset

The Child Mind Institute has launched the healthy brain network(HBN) initiative [1]. Within this initiative, they are creating a large biological databank starting from more than 10,000 people (children and adolescent). The databank stores a large amount of data related to behavioral and cognitive phenotyping, as well as multimodal brain imaging, electroencephalography (EEG), eye tracking, genetics, digital voice, and video samples. Data are accessible online.[17] The user may download MRI data, and then derive connectomes applying standard data processing pipeline.

5.4.2 The human connectome project

The human connectome project (HCP) [27] is a big project that aims to provide the community with insight into brains related to connectivity, functions, and variability among individuals. HCP is an effort of more than 5 years based on a data acquisition plan and a subsequent pipeline of analysis held by a consortium of investigators. The HCP focuses a cohort of 1200 subjects (twins and their nontwin siblings) using multiple imaging modalities (i.e., diffusion imaging, functional MRI, weighted MRI, electroencephalography, behavioral and genetic data.

Bringing together multiple resonance imaging modalities from different laboratories has been one of the significant challenges of the HCP. Therefore they developed a template pipeline for acquiring and storing data described in [9]. The pipeline is based on a set of minimal preprocessing pipelines that must be followed by all the participants to accomplish many low-level tasks. This allows the data interchange and, more important, the possibility of an easy comparison among different connectomes, reducing both storage and processing requirements.

5.4.3 The brain graph project and its database

Starting from data of the human connectome project, Kerepesi et al. [18] computed structural connectomes of 426 human subjects. For each individual, they used five different resolution scales, yielding (83, 129, 234, 463, and 1015 nodes) and many edge weights. All data are available in the GraphML language for download and authors also provide anatomically relevant annotations. Authors also offer for a subset of subjects the anatomical classification of subgraphs for some region of interest of the brain.

---

[17]http://fcon_1000.projects.nitrc.org/indi/cmi_healthy_brain_network/sharing_neuro.html.

Authors also offer the community a set of tools for processing connectomes through the GitHub interface.

5.5 General purpose repository of networks

Other than the biological network, an extensive collection of social networks are also available for network analysis research.

5.5.1 SNAP: the Stanford network analysis project

The Stanford network analysis platform (SNAP) [20], is a general-purpose system for analysis and management of networks that is freely available through a website http://snap.stanford.edu/. SNAP is implemented in different languages, such as C++ and Python. It offers over 140 different graph algorithms that can efficiently manipulate large graphs, calculate structural properties, generate regular and random graphs, and handle attributes and metadata on nodes and edges.

An important part of SNAP is the network dataset collection [19] that contains more than 80 different social and information real-world networks and datasets. Networks models many domains, such as biological networks, social networks, citation and collaboration networks, as well as Web and media networks.

Table 5.3 gives the types of datasets in the collection. The datasets are collected as part of our research in the past and, in that sense, represent typical graphs being analyzed. It gives the distribution of graph sizes in the collection. It can be observed that a vast majority of graphs are relatively small, with less than 100 million edges, thus can easily be analyzed in SNAP.

5.5.2 Network repository

NetworkRepository (NR)[18] [25] is a data repository for a network of different types (e.g., brain networks, social networks, etc.) available through a web-based platform. It stores more than 1000 various systems, and it also provides interactive visual analysis and an interactive graph analytics platform. Therefore the user of this database can manage, visualize networks, as well as analyze single systems or compare multiple networks using network statistics. NR also enables collaboration among users by allowing the users to discuss datasets, and to make a correction on data and analytics. The social aspect of NetworkRepository is a unique characteristic among others.

[18]networkrepository.com.

Table 5.3 Networks in publicly available repositories for benchmarking and analysis.

Network usage	Dataset	Available	#Nodes	#Edges
Social network	Karate network	UCI[a]	34	78
	American College Football		115	616
	Dolphin social network		62	159
	Books about US politics		105	441
	Les miserables		76	254
	Word adjacencies		112	425
	Arxiv general relativity (ca-GrQc)	SNAP[b]	5,242	14,496
	Arxiv high energy physics theory (ca-HepTh)		9,877	25,998
	Social circles from Facebook		4,039	88,234
	Arxiv condensed matter (ca-CondMat)		23,133	93,497

[a] https://networkdata.ics.uci.edu/.
[b] https://snap.stanford.edu/data/.

5.6 Summary

Biological networks produced by experimental platforms are stored into different databases. Such data are the essential building block for all the subsequent analysis tasks. Therefore the characteristics of such data storage systems may dramatically impact the performances of the subsequent steps.

Initially, many different groups used classic storage systems based on the relational model or simply flat files. Despite the simplicity, it has been shown that both relational and flat file model present many limitations related to speed and querying capabilities. Therefore, the growth of No-SQL databases has offered the possibility to use such systems even in network storage. Here we surveyed main approaches and some related experiences. More specifically we discussed and reported few network data sources available for analyzing genetic, proteomics and brain connectome networks. We even discussed few large scale social network data sources which may help other network researchers too.

References

1. Lindsay M. Alexander, J. Escalera, et al., An open resource for transdiagnostic research in pediatric mental health and learning disorders, Scientific Data 4 (2017) 170181.

2. C. Alfarano, C.E. Andrade, et al., The biomolecular interaction network database and related tools 2005 update, Nucleic Acids Research 33 (Database issue) (January 2005) 418–424.

3. Renzo Angles, Marcelo Arenas, Pablo Barceló, Aidan Hogan, Juan Reutter, Domagoj Vrgoč, Foundations of modern query languages for graph databases, ACM Computing Surveys 50 (5) (2017) 68.

4. Renzo Angles, Claudio Gutierrez, Survey of graph database models, ACM Computing Surveys 40 (1) (2008) 1–39.

5. K.R. Brown, I. Jurisica, Online predicted human interaction database, Bioinformatics 21 (9) (May 2005) 2076–2082.

6. Mario Cannataro, Pietro H. Guzzi, Pierangelo Veltri, Protein-to-protein interactions: technologies, databases, and algorithms, ACM Computing Surveys 43 (1) (2010) 1.

7. Yves Deville, David Gilbert, Jacques van Helden, Shoshana J. Wodak, An overview of data models for the analysis of biochemical pathways, Briefings in Bioinformatics 4 (3) (2003) 246–259.

8. Nadime Francis, Alastair Green, Paolo Guagliardo, et al., Cypher: an evolving query language for property graphs, in: Proceedings of the 2018 International Conference on Management of Data, SIGMOD '18, ACM, New York, NY, USA, 2018, pp. 1433–1445.

9. Matthew F. Glasser, Stamatios N. Sotiropoulos, J.A. Wilson, et al., The minimal preprocessing pipelines for the human connectome project, NeuroImage 80 (2013) 105–124.

10. Heonjong Han, Jae-Won Cho, Sangyoung Lee, Ayoung Yun, Hyojin Kim, Dasom Bae, Sunmo Yang, Chan Yeong Kim, Muyoung Lee, Eunbeen Kim, et al., Trrust v2: an expanded reference database of human and mouse transcriptional regulatory interactions, Nucleic Acids Research 46 (D1) (2017) D380–D386.

11. Christian Theil Have, Lars Juhl Jensen, Are graph databases ready for bioinformatics?, Bioinformatics 29 (24) (2013) 3107.

12. H. Hermjakob, L. Montecchi-Palazzi, et al., Intact: an open source molecular interaction database, Nucleic Acids Research 32 (Database issue) (2004) 452–455.

13. H. Hermjakob, et al., The hupo psi's molecular interaction format– a community standard for the representation of protein interaction data, Nature Biotechnology 22 (2) (2004) 177–183.

14. Sheng-Da Hsu, Feng-Mao Lin, W.Y. Wu, et al., mirtarbase: a database curates experimentally validated microrna-target interactions, Nucleic Acids Research 39 (suppl 1) (2010) D163–D169.

15. Tao-Wei Huang, An-Chi Tien, Wen-Shien Huang, Yuan-Chii G. Lee, Chin-Lin Peng, Huei-Hun Tseng, Cheng-Yan Kao, Chi-Ying F. Huang, Point: a database for the prediction of protein-protein interactions based on the orthologous interactome, Bioinformatics 20 (17) (2004) 3273–3276.

16. Minoru Kanehisa, et al., The kegg database, in: Novartis Foundation Symposium, Wiley Online Library, 2002, pp. 91–100.

17. Minoru Kanehisa, Susumu Goto, Miho Furumichi, Mao Tanabe, Mika Hirakawa, Kegg for representation and analysis of molecular networks involving diseases and drugs, Nucleic Acids Research 38 (suppl 1) (2009) D355–D360.

18. Csaba Kerepesi, Balazs Szalkai, Balint Varga, Vince Grolmusz, The braingraph.org database of high resolution structural connectomes and the brain graph tools, Cognitive Neurodynamics 11 (5) (2017) 483–486.

19. Jure Leskovec, Andrej Krevl, SNAP datasets: Stanford large network dataset collection, http://snap.stanford.edu/data, June 2014.

20. Jure Leskovec, Rok Sosič, Snap: a general-purpose network analysis and graph-mining library, ACM Transactions on Intelligent Systems and Technology 8 (1) (2016) 1.

21. Zhi-Ping Liu, Canglin Wu, Hongyu Miao, Hulin Wu, Regnetwork: an integrated database of transcriptional and post-transcriptional regulatory networks in human and mouse, Database (2015) 2015.

22. Antonio Messina, Antonino Fiannaca, Laura La Paglia, Massimo La Rosa, Alfonso Urso, Biograph: a web application and a graph database for querying and analyzing bioinformatics resources, BMC Systems Biology 12 (5) (2018) 98.

23. Pablo Pareja-Tobes, Raquel Tobes, Marina Manrique, Eduardo Pareja, Eduardo Pareja-Tobes, Bio4j: a high-performance cloud-enabled graph-based data platform, 2015.

24. Marko A. Rodriguez, The gremlin graph traversal machine and language (invited talk), in: Proceedings of the 15th Symposium on Database Programming Languages, ACM, 2015, pp. 1–10.

25. Ryan A. Rossi, Nesreen K. Ahmed, The network data repository with interactive graph analytics and visualization, in: AAAI, 2015.

26. Chris Stark, Bobby-Joe Breitkreutz, Teresa Reguly, Lorrie Boucher, Ashton Breitkreutz, Mike Tyers, Biogrid: a general repository for interaction datasets, Nucleic Acids Research 34 (suppl 1) (2006) D535–D539.

27. David C. Van Essen, Kamil Ugurbil, E. Auerbach, D. Barch, et al., The human connectome project: a data acquisition perspective, NeuroImage 62 (4) (2012) 2222–2231.

28. Ioannis Xenarios, Lukasz Salwinski, Xiaoqun Joyce Duan, P. Higney, et al., Dip, the database of interacting proteins: a research tool for studying cellular networks of protein interactions, Nucleic Acids Research 30 (1) (2002) 303–305.

29. Fang Zhao, Zhenyu Xuan, Lihua Liu, Michael Q. Zhang, Tred: a transcriptional regulatory element database and a platform for in silico gene regulation studies, Nucleic Acids Research 33 (suppl 1) (2005) D103–D107.

30. Pablo Pareja-Tobes, Raquel Tobes, Marina Manrique, Eduardo Pareja, Eduardo Pareja-Tobes, Bio4j: a high-performance cloud-enabled graph-based data platform, bioRxiv, https://doi.org/10.1101/016758, 2015.

6

Gene expression networks: inference and analysis

Contents

The influences of one gene in the activity of another gene is important, yet to identify experimentally in a wet lab environment. Hence, computational tools and methods are playing pivotal role in inferring near optimal network structures, which can further be confirmed using wet lab experimentation. The modeling (reconstruction) of a Gene Regulatory Network (GRN) based on experimental data is also called reverse engineering or network inference. Inference of gene networks can be defined as the process of identifying gene interactions from experimental data through computational analysis. We introduce here the gene-gene inter-

Biological Network Analysis. https://doi.org/10.1016/B978-0-12-819350-1.00012-8

action networks and various inferencing methods for reverse engineering of such networks. We divide our discussion into three major steps. The expression data generation, inferencing gene regulatory networks and post-inference analysis of the inferred networks. We also discuss few tools or softwares available for inference and visualization of the gene networks.

6.1 Expression network and analysis: the workflow

Gene expression has become very essential in system-level understanding of behavior of genes. High-throughput microarray and next generation sequencing (NGS) technology makes available a large repository of expression data. It allow us to study the dynamic behavior of a gene inside a cell. Reverse engineering is a promising area of research in systems biology; it aims to recreate the cellular system for better understanding of biological mechanism. The development of a suitable reverse engineering method is important to get insight into the gene-gene relationships. Study of such data may enable us to address various issues, such as how a gene participates in a cellular process; what are the activities of different genes; in which cell and under which conditions, do the genes become active; how the activity of a gene is influenced by various diseases or drugs, and how genes contribute to diseases. One of the major goals in analyzing expression data is to determine how the expression of any particular gene may affect the expression of other genes or how one gene regulates another gene. Gene-gene relationships can be described through biological pathways, which can be represented as networks, broadly classified [66] as *metabolic pathways, signal transduction pathways,* and *gene regulatory networks.* The most preliminary network is the gene coexpression network, which describes certain association among genes.

Genes that affect one another may belong to the same gene network. A gene network is a set of related genes, where expression of one gene may influence the other gene's activity. A group of co-regulated genes may form gene clusters that can encode proteins, which interact amongst themselves and take part in common biological processes. *In silico* reconstruction of such biological networks is essential for exploring regulatory mechanisms and is useful in better understanding of the cellular environment to investigate complex interactions [43]. In an organism, coexpression of genes depend on their sharing of the regulatory mechanism. It has been observed that genes with similar expression profiles are very likely to be regulators of one another, or be regulated by some other common parent gene [26]. Another major goal of expression data analysis is to determine what genes are over-expressed or underexpressed as a result of certain biological conditions, such as, what genes are expressed in diseased cells that are not expressed in normal cells. Recently, it has been observed that a small set of genes are coregulated and coexpressed under certain conditions, and their behavior being almost inactive for rest of the conditions. Discovering a group of genes with similar or inverted expression profiles has been employed to identify coexpressed group of genes (termed as modules), as well as to extract gene interactions or gene regulatory networks [66]. Isolating well-connected genes within the module using various topological analysis of the subnetwork may help in identifying disease-related biomarkers or essential disease genes. Once important genes can be identified, further investigation in identifying small chemical molecule modulator for binding overexpressed key genes in disease condition may produce effective drug target. At the end of the day, all the tasks performed *in silico* should be verified biologically in a wet lab environment before final consideration. A possible workflow of overall gene expression inference and analysis for disease diagnosis and biomarker identification is depicted in Fig. 6.1. It is important to mention here that the steps shown in the figure may not be fixed and subject to variation, depending on the methodology adopted.

6.2 Basics of gene expression

Genes are nothing but regions of the DNA, and act as a repository of biological information, which is necessary to build and maintain an organism's cells. It includes construction and regulation of proteins, as well as other molecules that ultimately determine the growth and functioning of the living organism and transfer genetic traits to next generation. This is termed as the central dogma of molecular biology. Entire DNA sequence of an organism do not play active role in cellular activities. In the case of the human genome, only 2–3% of the whole human DNA are functional. The functional part or the coding part of DNA is only responsible for protein synthesis. The rest of the DNA consists of noncoding regions, and does not encode for any protein. This DNA is sometimes referred to as "junk-DNA" or noncoding DNA. Recent research reveals that junk-DNA plays critical roles in controlling how cells, organs, and other tissues behave. The coding part of DNA, gene, decides the type of protein that will be produced within a cell. Protein synthesis takes place within the cell through

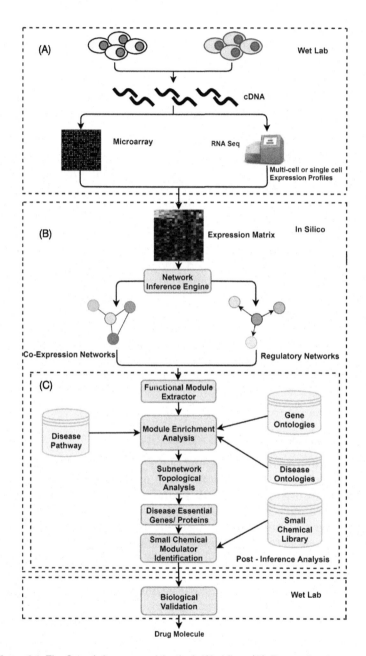

Figure 6.1. The Gene Inference and Analysis Workflow. (A): Expression data generation platform. (B): In-silico inference of networks. (C): Post inference network analysis.

the process of transcription and translation. In the *transcription* phase, a molecular complex called RNA polymerase-II creates a copy of a gene from the DNA to messenger RNA (mRNA) inside the nucleus. The mRNA travels from nucleus to the cytoplasm for protein synthesis, where it then binds with ribosome. Ribosome is a complex molecule based on ribosomal RNA (rRNA) and proteins. At the ribosome, mRNA is used as a blueprint for the production of a protein; this process is called *translation*. The mRNA moves along the protein synthesis site, i.e., ribosomes, with a set of three-nucleotides called *codons*. Transfer RNA (tRNA) provides a compatible *anticodon* and is hybridized onto the mRNA. Finally, the amino acids bound to the RNA form a polypeptide chain. This process continues until the translation process reaches a stop codon, which terminates the polypeptide synthesis. The entire process is called **gene expression**.

Traditional experimentation systems in molecular biology are capable of studying only a few genes in a single experiment. Moreover, for a traditional method, it is difficult to capture the dynamic behavior or the activities of a gene that is going on inside a cell. Advent of high-throughput technology makes it possible to generate expression profiles of large chunk of genes in different biological environment and time course. DNA microarray and most recently developed next generation sequencing (NGS) technology provides a convenient and effective platform for monitoring activity of thousands of genes simultaneously.

6.2.1 Microarray data generation

Microarray is an indispensable technology in molecular biology that helps in assessing expression of a large number of genes under multiple conditions, such as time-series, tissue samples (e.g., normal versus cancerous tissues), and experimental conditions. With the help of microarray experiments, one can monitor, simultaneously, the expression levels of several genes at a genome scale. To gain better understanding of a gene and its behavior inside a cell, various patterns can be derived by analyzing the change in expression of the genes.

There are five major steps in performing a typical microarray experiment[1]:

- mRNA isolation: This step includes the isolation of RNA from the cells. Degradation of RNA in this step is monitored by RNA electrophoresis or a bioanalyzer using RNA integrity number (RIN) as a benchmark.

[1]http://grf.lshtm.ac.uk/microarrayoverview.htm.

- cDNA synthesis: At the advent of this stage, RNA controls are added, which are used for controlling the synthesis of complementary DNA (cDNA). This synthesis is carried out with the help of oligo-dT or random primers. This step enables the process of reverse transcription.

- Amplification and labeling: cDNA, synthesized in the prior step is amplified using in vitro transcription. The main motive of this step is to acquire a cRNA, comprising biotinylated C and U nucleotides, required in the later steps.

- cRNA fragmentation: In this step, cRNAs obtained from the previous step is cut into many fragments. The next step of hybridization is applied to these fragments after it is transferred onto the microarray chip.

- Hybridization: This is a step, which removes any kind of anomaly found in the previous step. This step is an amalgamation of many steps depending on reaction condition and structural properties, which would play a significant role in the outcome of the cDNA molecules. This step is carried out for controlling the consistency of the overall microarray performance.

- Staining: This step is applied after washing the outcome from the prior step, which removes those cRNA bound to the microarray surface. Staining helps in stabilizing the microarray data.

- Image acquisition and data analysis: The concluding step of microarray data preparation is image acquisition. The slide obtained from the previous step is dried, upon which a laser scanner is used to check how much cDNA is bound to a target space. Depending upon the amount of cDNA bound to the target space, composition and representation of mRNA composition are determined. Software used for microarray analysis often represents upregulation (a gene compared the to control) by a green spot, downregulation (a gene in the experimental sample) by a red spot, and yellow to represent equal abundance in both experimental and control samples. In the data analysis phase, the relative expression levels of the genes in the sample and in the controlled populations can be estimated from the fluorescence intensities and color for each spot. Based on the amount of probe hybridized to each target spot, information is gained about the specific mRNA composition and the representative in the sample. The logarithm of the ratio of raw red/green fluorescence intensities are taken to convert them into log intensities. The data is normalized before use for the inference and analysis.

6.2.2 RNA-seq read counts

RNA sequencing (RNASeq) [70] is an alternative to microarray and more effective high-throughput technology, introduced recently to capture gene expression profile. It brings remarkable improvements in expression data analysis. Expression of a gene can be quantified by the number of reads mapped to a particular RNA, termed as RNA-seq count. The RNA-seq counting uses next-generation sequencing (NGS) to extract amount of RNA contents in a target sample.

The principle behind **next generation sequencing (NGS)** is the capillary electrophoresis. At first, the genomic strands are fragmented, and the bases are identified in each fragment by ligation with custom linkers or template strand. The NGS uses array-based sequencing method to process millions of reactions in parallel, in very high speed and at a reduced cost. There are three general steps involved in NGS: (i) library preparation, (ii) amplification, and (iii) sequencing. Before sequencing, the isolated and purified RNA must be converted to double-stranded complementary DNA (cDNA). The candidate RNA to be sequenced first converted into cDNA fragments or cDNA library. Sequencing platform specific adapters are then added to each end of the fragments. Adding adapters and amplification of DNA to make a cDNA are the steps in library preparation. Finally, the cDNA library is sequenced using NGS. RNA-seq offers higher resolution for low-abundance transcripts, and helps in identifying tissue-specific expression better. In addition, the expression of different splice variants [54,68] can be differentiated well by RNA-seq data.

The rationale behind using NGS or microarray is that these technologies enable the investigation of the level of the activity of each gene through a quantitative analysis. For the last few decades, microarray remains as the only technology to capture the expression levels. However, with the availability of high-throughput NGS technologies, it is now possible to analyze whole genome sequence, including whole transcriptome analysis. Due to hybridization, the array is susceptible to background noise. RNA-seq technology produces read counts, which quantify the expression of a larger dynamic range with low noise rate. RNA-seq technology is capable to detect differentially expressed genes, even genes with low expression level.

6.2.3 Single-cell transcriptomics

The advancement of RNA sequencing technology helps in generating high resolution single-cell gene expression data [46]. Single-cell transcriptomics revolutionized the way of understanding the cellular mechanism of interactions by understanding the heterogeneity of intracellular activities with varying cell types.

6.2.4 RNA-seq vs. microarray

RNA-seq counts has been preferred over microarray for capturing gene expression due to several distinct advantages [29]. First, unlike preselected probes in microarray, the RNA-seq is able to identify novel genes. Second, RNA-seq is capable to capture expression at the gene, exon, transcript, and coding DNA sequence (CDS) levels, whereas resolution of microarray is limited to the level of genes. Finally, RNA-seq can also extract alternative splicing.

The level of expression once captured is mapped into two dimensional matrix for further analysis of the expression data.

6.2.5 Expression data matrix

An expression profile (of a gene or a sample) can be represented in vector space [28]. For example, an expression profile of a gene can be considered a vector in n dimensional space (where n is the number of conditions), and an expression profile of a sample with m genes can be considered a vector in m dimensional space (where m is the number of genes). In the example given below, the gene expression matrix X with m genes across n conditions is an $m \times n$ matrix, where the expression values for gene i in condition j is denoted as x_{ij}:

$$X = \begin{bmatrix} x_{1,1} & x_{1,2} & \cdots & x_{1,n} \\ x_{2,1} & x_{2,2} & \cdots & x_{2,n} \\ \vdots & \vdots & \ddots & \vdots \\ x_{m,1} & x_{m,2} & \cdots & x_{m,n} \end{bmatrix}. \tag{6.1}$$

Formally, we can define gene expression data as follows:

Definition 6.2.1 (Gene expression data). Let $G = \{g_1, g_2, \cdots, g_m\}$ be a set of m genes, and let $T = \{t_1, t_2, \cdots, t_n\}$ be the set of n conditions or time points (or RNA sequence reads). The gene expression dataset X can be represented as an $m \times n$ matrix, i.e., $X_{m \times n}$, where each entry $x_{i,j}$ in the matrix corresponds to the logarithm of the relative abundance of mRNA of a gene or RNA counts.

The expression profile of a g_i can be represented as a row vector:

$$g_i = \begin{bmatrix} x_{i,1} & x_{i,2} & x_{i,3} & \cdots & x_{i,n} \end{bmatrix}. \tag{6.2}$$

The expression profile of a sample j can be represented as a column vector:

$$g_i = \begin{bmatrix} x_{1,j} \\ x_{2,j} \\ x_{3,j} \\ \dots \\ x_{m,j} \end{bmatrix}.$$

A subset of real gene expression data from a Homo sapiens (GDS825) derived from NCBI[2] is shown in Table 6.1. Three different genes (row) and their relative expression values with respect to four different samples or time course (column) are shown in the table.

Table 6.1 Sample expression data matrix.

ORF	C1	C2	C3	C4
GALNT5	−3.474	−3.837	−4.644	−5.059
APOE	−2	−1.943	−1.786	−1.737
IDH3B	1.449	1.299	0.993	0.832

6.3 Inference of expression networks

The limitation of experimental wet lab technologies is that it cannot measure mutual influences among all genes from one organizm's genome simultaneously, therefore computational methods are applied to infer and reveal mutual gene interactions. Analysis and interpretation of the relationships in biological networks is becoming a major research area of interest in modern computational biology, and its translation to genomic medicine. Several difficulties arise when dealing with large regulatory networks involving thousands of genes/protein interactions. One severe bottleneck is the visual analysis and the interpretation of these regulatory networks. This has led the scientific community, computer engineers, statisticians, and biologist to come together to develop new methodology and algorithm to address these issues by developing open-source reconstruction and visualization tools.

Once the expression is available, it is further preprocessed [57] to remove experimental noise (if persists). Normalization of ex-

[2] www.ncbi.nlm.nih.gov.

pression values is another important step before considering the expression matrix fit for further analysis. Expression matrix records relative abundance of mRNA levels in a target sample and scale of levels may vary with experimentation environment, and hence normalization is important. Preprocessed expression data is used directly in differential expression analysis, coexpressed gene clustering or biclustering [55]. However, among them the most important task is to infer computationally the gene interactions, which ultimately forms a graph or network of gene expression levels. Two types of graph are usually inferred using expression data, such as directed and undirected. It is challenging to infer the directed network, termed as causal or gene regulatory network, among genes or proteins, due to lack of sufficient information in the available data sources. Such data are limited in inferring coexpression networks only, as true regulatory information is missing. Inferring undirected graph from expression matrix is relatively easy in comparison to directed graph. The guilt-by-association in the form of coexpression may be measured in terms of statistical correlations, mutual information, or some other similarity measures between the genes' expressions. The count for coexpression network inference methods is much higher than regulatory inference methods. If we consider a simplistic view of the network inference, it is a task of computationally converting expression matrix into another matrix, called adjacency matrix. The target is to generate adjacency matrix as close as possible to the actual interactions. However, interestingly, known interactions derived based purely on expression data is missing. Hence, validating any method about its quality of prediction is also equally challenging.

6.3.1 Inferring causal gene regulatory networks

The graph theoretic formalism of expression network is the common and simplistic way of representing it.

Definition 6.3.1. Gene regulatory network (GRN): A GRN is a graph, $\mathcal{G} = (\mathcal{V}, \mathcal{E})$, where $\mathcal{V}$ is the set of all genes in the network and $\mathcal{E}$ is the set of edges between a pair of genes, say $(v_i, v_j) \in \mathcal{V}$, representing a strong biomolecular interaction between the two genes.

A directed edge from node v_i to v_j ($v_i \rightarrow v_j$) indicates that a causal effect exists from node v_i to v_j. Causality provides the direction of influence between the two genes, called cause and effect, respectively. The directed edges in a GRN correspond to causal influences between gene activities (nodes). These may include regulation of transcription by transcription factors, but also less intuitive causal effects between genes, involving signal transduction or metabolism.

A causal effect may be direct or indirect [10]. A gene may influence activities of other genes or gene products directly. It is also possible that a gene may influence activities of other genes, or itself, by coding a transcription factor (TF) that in turn regulates another gene, or itself. Some possible causal relationships within a GRN are shown in Fig. 6.2, which illustrates the following types of causal relationships [2]:

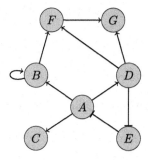

Figure 6.2. Possible causal relationship between the nodes in a GRN.

1. **One-to-many:** A gene can influence the activities of more than one gene. Gene A regulates positively the expression of genes B, C, and D.
2. **Many-to-one:** A gene's activity may be influenced by more than one gene (relationships among B, D, and F).
3. **Feedback loop:** In a feedback loop, a gene influences the activities of some of its ancestors in its regulatory pathways. For example, E regulates the expression of gene A negatively.
4. **Feed-forward loop:** The regulation among D, F, and H is called a feed-forward regulatory structure, where H is directly and indirectly influenced by D.
5. **Self-loop:** A gene can influence its own activity (node B).
6. **Inhibition:** A gene may inhibit activity of another gene (D inhibits E). Inhibition or negative regulation may take place in many-to-one, one-to-many, and different loop structures.

Extraction of cause and effect relations in GRNs will not only create a blueprint for the complexity of cellular mechanisms, but may also open up research, involving probable causes of diseases, such as cancer and AIDS, as well as neurodegenerative diseases, such as Alzheimer's or Parkinsons [1,72]. Hence, the computational inference of causal edges in GRNs is a very important, although a challenging task.

In what follows, we discuss few causal network inference methods.

6.3.2 Directed network inference methods

LBN LBN [40] is a novel local Bayesian network (LBN) algorithm to reconstruct GRNs from gene expression data that is proposed to overcome the limitations of Bayesian networks (BN), mutual information (MI) and conditional mutual information (CMI) by making use of their advantages, i.e., infer the directed network with less false-positive edges and with high computational efficiency. LBN algorithm mainly consists of five distinct elements: first, CMI is employed to construct an initial network, which is decomposed into a series of smaller subnetworks according to the nearest relationship among genes in the network with k-nearest neighbor (kNN) method; second, BN method is used to identify their regulatory relationships with directions, generating a series of local BNs, which are integrated into a candidate GRN; third, CMI is applied to remove the false positive edges, forming a tentative GRN; fourth, the tentative GRN is further decomposed into a series of smaller subnetworks or local networks, in which BN method is implemented to delete the false regulatory relationships, and fifth, the final network or GRN is inferred by iteratively performing BN and CMI with kNN decomposition until the topological structure of the tentative network does not change. LBN method can infer most of the true regulatory relationships between genes, and the effectiveness and efficiency of LBN method is verified on the real gene expression data.

DELDBN Differential equation-based local dynamic Bayesian network (DELDBN) [38] is a hybrid algorithm integrating ordinary differential equation (ODE) models with dynamic Bayesian network (DBN) analysis. ODE models are used to represent a parametric version of mechanistic gene regulations, and capture the underlying data generation process. DBN analysis, using local causality discovery algorithm, is then performed to connect the left-hand side (transcription rate of a gene) to the right-hand side of an ODE (expression level of regulatory genes).

GENIE3 GEne network inference with ensemble of trees [31] is based on variable selection with ensembles of regression trees. It was the best performer in the DREAM4 (Dialogue on Reverse Engineering Assessment and Methods)[3] in silico network challenge. GENIE3 decomposes the problem of inferring a network of size

[3]http://dreamchallenges.org/.

p into p different feature selection problems, where the goal is to identify the regulators of one of the genes of the network using tree-based ensemble methods. It does not make any assumption about the nature of gene regulation, and is able to deal with combinatorial regulations and nonlinearity. GENIE3 works well even with a large number of genes. It is computationally fast, and scalable.

JUMP3   JUMP3 [59] is a hybrid approach to bridge the gap between model-free and model-based methods. It uses a formal stochastic differential equation to model each gene expression, but reconstructs the GRN topology with a nonparametric method based on decision trees. The idea of Jump3 is to learn about each target gene, a tree-based model that will predict the promoter state at any time point from the expression levels of the other genes (or a set of candidate regulators) at the same time point. On applying it on a real gene expression dataset related to murine macrophages to retrieve regulatory interactions that are involved in the immune response, the predicted network found to be highly modular, with a few transcription factors acting as hub nodes and each one regulating a large number of target genes. The hub transcription factors are found to be biologically relevant, comprising some interferon genes, with one gene known to be associated with cytomegalovirus infection and several cancer-associated genes.

SWING   Sliding window inference for network generation (SWING) [23] is a generalized framework that incorporates multivariate Granger causality [27] to infer network structure from time-series data. SWING generates windowed models that simultaneously evaluate multiple upstream regulators at several potential time delays. SWING embeds existing multivariate methods, both linear and nonlinear, into a Granger causal framework that concurrently considers multiple time delays to infer causal regulators for each node. SWING also uses sliding windows to create many sensitive, but noisy, inference models that are aggregated into a more stable and accurate network. SWING is validated on several *in silico* time-series datasets and existing *in vitro* datasets with corresponding gold standard networks. SWING exhibits a statistically significant increase in AUROC and AUPR for many of the 10-node networks, and across all of the 100-node networks. SWING outperforms the other network inference algorithms, but is limited by computational expense.

PGC   The Prophetic Granger Causality (PGC) [13] introduces a regularized regression framework inspired by granger causality that appropriately handles irregularly spaced time intervals. In contrast to the traditional granger causality approach that uses only past observations, PGC introduces a "prophetic" extension. It includes future observations, to consider interaction evidence from the perspective of both the regulator and the target. The PGC method was a top-performer in the DREAM8 competition, producing interactions with a higher likelihood of representing causal connections compared to other methods. DREAM8 evaluated methods using wet lab experiments conducted after all algorithm predictions were collected. This method outperformed 73 other submissions in the final scoring round, which lends credibility to its strength.

TIGRESS TIGRESS [30] (Trustful Inference of Gene REgulation with Stability Selection) was one of the top GRN inference methods in the DREAM5 gene network inference challenge. It formulates the GRN inference problem as a sparse regression problem using the feature selection method—Least Angle Regression (LARS) [20] and stability selection [49] method. On comparison with other methods it is found that it achieves a similar performance to that of GENIE3 in synthetic networks, but do not perform as well in real networks.

SINCERITIES SINgle CEll regularized inference using TImestamped Expression profileS SINCERITIES [53] infers GRNs from single-cell transcriptional profiles by recovering directed regulatory relationships among genes using regularized linear regression (ridge regression) and temporal changes in the distributions of gene expressions. It is validated using in silico time-stamped single cell expression data and single cell transcriptional profiles. Results show that SINCERITIES could provide accurate GRN predictions significantly better than other GRN inference algorithms, such as GENIE3, and JUMP3. It also has a low computational complexity and is amenable to problems of extremely large dimensionality.

MIBNI Mutual information-based Boolean network inference (MIBNI) algorithm [7] first employs a mutual information-based feature selection (MIFS) method to approximate the mutual information between a target gene and a set of candidate regulatory genes, which reduces the computational cost. MIBNI then utilizes a SWAP subroutine, wherein a gene in the set of regulatory genes selected by MIFS is iteratively swapped with another gene in the set of unselected genes. MIBNI repeats the MIFS and SWAP subroutines by turns until the desired number of most informative regulatory genes with the highest dynamics accuracy is found. MIBNI exhibits higher precision, recall, and structural accuracy values when number of incoming links is more than 1. MIBNI also shows significantly higher dynamics accuracy which increases as the number of incoming links increase.

GABNI GABNI [8] is a Genetic Algorithm-based Boolean Network Inference method. First it applies MIBNI. When MIBNI fails to find an optimal solution, a genetic algorithm (GA) is applied to search an optimal set of regulatory genes in a wider solution space. Basically, a typical GA framework is modified to efficiently reduce a search space. The performance of GABNI with both the artificial and the real gene expression datasets is validated through comparisons with existing methods like MIBNI, GENIE3, and TIGRESS. Results demonstrate that GABNI significantly outperformed other algorithms in both structural and dynamics accuracies. The structural performance of GABNI is more effective than the other methods as the incoming links increases. The dynamic accuracy of GABNI is 1.0 (perfect accuracy) for all network sizes, whereas that of MIBNI, GENIE3, and TIGRESS methods decreased as the network size increased. GIBNI, however, finds a high-quality solution by sacrificing the search time.

SCODE SCODE [47] is based on linear ODEs, more specifically, on the transformation of linear ODEs and linear regression with significantly small time complexity. SCODE is validated using three scRNA-seq datasets during differentiation, and it shows that SCODE can successfully optimize ODEs to reconstruct observed expression dynamics. The AUC values of SCODE are higher than those of other methods in almost of all cases. The runtime of SCODE is significantly smaller than that of JUMP3, and GENIE3.

6.3.3 Coexpression network inference

A gene coexpression network is an undirected graph, where the nodes correspond to genes or gene activities, and undirected edges between genes represent significant coexpression relationships [18]. In a coexpression network, two genes are connected by an undirected edge if their activities have significant influence over a series of gene expression measurements. Compared to regulatory networks, a gene coexpression network does not attempt to draw direct causal relationships among the participating genes in the form of directed edges. Coexpression network analysis plays vital role in inferring relationship among biological processes [36].

A gene coexpression network is an undirected graph, and thus can use mathematical and computational tools associated with graphs.

Definition 6.3.2 (Gene coexpression network). A gene coexpression network is a graph $\mathcal{T} = (\mathcal{G}, \mathcal{E})$, where $\mathcal{G}$ denotes the set of N genes (nodes) $\{G_1, G_2, \cdots, G_N\}$ participating in a common gene product formation process or biological process, and $\mathcal{E}$ is the set of edges $\{e_1, e_2, \cdots, e_m\}$ that correspond to the similar expression levels among the genes across different samples, or times. An arc between two genes may contain weight signifying their relative proximity.

Few of the popular proximity measures that are used commonly includes correlation, mutual information, etc. The available coexpression network inference varies greatly, because of the association or proximity measures they use.

6.3.4 Undirected network inference methods

ARACNE   Algorithm for the Reconstruction of Accurate Cellular NEtworks (ARACNE) [45] is an information-theoretic algorithm that identifies candidate interactions by estimating pairwise mutual information between a pair of gene expression profiles. Initial network is then pruned by removing indirect candidate interaction using data processing inequality (DPI). ARACNE's performance is assessed by reconstructing a class of synthetic networks and a human B lymphocyte genetic network from gene expressions profile. The values of precision and recall for ARACNE are consistently better than the other tested methods. For any reasonable precision (i.e., > 40%), ARACNE has a significantly higher recall than the other methods, and its precision reaches 100% at significant recall levels. In a biological context, the algorithm infers bona fide transcriptional targets in a mammalian gene network. However, apart from identifying direct transcriptional interactions, ARACNE requires more research to precisely characterize other types of interactions.

CLR   The Context Likelihood of Relatedness (CLR) algorithm [22] is an extension of the relevance network [12] class of algorithms, but applies an adaptive background correction step to eliminate false correlations and indirect influences to overcome the limitations of relevance networks. The mutual information between regulators and their potential target genes is computed, and then CLR calculates the statistical likelihood of each mutual information value within its network context. Several versions of relevance networks, ARACNe, Bayesian networks, and regression networks are compared with CLR. CLR is not only the top-performing inference method, but also predicted 1079 regulatory interactions at a 60% true positive rate, of which 338 are in the previously known network, and 741 are novel predictions. The predicted interactions are tested for three transcription factors with chromatin immunoprecipitation, confirming 21 novel interactions and verified with the RegulonDB[4] an online database for regulatory network of E. coli. Although CLR performe better than ARACNE and the other algorithms, and recovered more interactions with higher precision, it accounted for only a fraction of known interactions in *E. coli*.

MRNET   MRNET [51] is another information-theoretic method inspired by a feature selection technique called the maximum relevance/minimum redundancy (MRMR) algorithm. MRNET formulates the network inference problem as a series of input/output supervised gene selection procedures, and then adopts the MRMR principle to perform the gene selection for each supervised gene selection procedure. MRNET is benchmarked against three information-theoretic network inference methods: relevance networks (RELNET), CLR, and ARACNE. The comparison relies on thirty artificial microarray datasets synthesized by two public-domain generators. Results show that MRNET is competitive with the benchmarked information-theoretic methods, and that MRNET and CLR are the two best techniques when the nonparametric Miller–Madow estimator is used.

C3NET   C3NET [3] is an unsupervised method to identify undirected networks that works by identifying a significant maximum mutual information network in a way that two genes are connected with each other only if their shared significant mutual information value is at least (for one of these two genes) maximal with respect to all other genes. It is benchmarked with ARACNE, MRNET, RELNET, and CLR [25] for simulated as well as expression data from E. coli. C3NET gives better results than all other inference methods and, in addition, has the lowest computational complexity.

GeCON   GeCON [56] is an expression pattern-based method to extract Gene CO-expression network from microarray data. GeCON uses pattern matching to capture pairwise similarity considering both positive and negative regulation as coregulation.

[4]http://regulondb.ccg.unam.mx/.

The similarity between two expression patterns is computed using a support-based approach. It stresses the development of a computationally effective network reconstruction technique, and therefore computes the similarity between a pair of genes using a fast one-pass support count based approach, wherein strong support between a pair of genes represents strong association between them. Gene pairs showing high support, i.e., high pattern similarity are then used to construct a gene coexpression network. GeCON is applied on several synthetic and real expression datasets. The results are assessed using real datasets by evaluating the network modules extracted from the network against biologically significant gene ontology (GO) terms associated with a group. It is benchmarked with ARACNE, CLR, and MRNET. GeCON outperforms all other algorithms in terms of network prediction and computation time. It is also faster than ARACNE. Results also show that network modules extracted have high biological significance.

PIDC   PIDC [14] uses partial information decomposition (PID) to identify regulatory relationships between genes. The algorithm identifies putative functional relationships between genes based on pairwise MI, combined with information about the local network context of each gene. The performance evaluation of the algorithm along with other inference algorithms, such as ARACNE, CLR, MI (relevance networks), MRNET, and the PID-based algorithm, shows that the higher-order information captured by PIDC allows it to perform better than other pairwise mutual information-based algorithms in recovering true relationships present in simulated data.

WGCNA   Weighted Gene Coexpression Network Analysis (WGCNA) [73] infers correlation-based weighted coexpression network. The network is defined using a gene coexpression similarity measure that is denoted as s_{ij} for a pair of genes i and j. To define a signed coexpression measure between gene expression profiles x_i and x_j, a simple transformation of the following correlation is used, where the signed similarity s_{ij}^{signed} takes on a value between 0 and 1:

$$s_{ij}^{signed} = 0.5 + 0.5cor(x_i, x_j). \qquad (6.3)$$

An adjacency matrix $A = [a_{ij}]$ is subsequently obtained by thresholding the coexpression similarity matrix $S = [s_{ij}]$ to express the measure on how strongly genes are interconnected. Soft thresholding is applied instead of hard thresholding, so that the continuous nature of the coexpression information can be preserved. The connection strength is assessed using the following power function:

$$a_{ij} = (s_{ij})^\beta, \tag{6.4}$$

where the power β is the soft thresholding parameter.

GBM   Graphical Gaussian models (GGMs) [60] represent multivariate dependencies in biomolecular networks by means of partial correlation and can be implemented using the GeneNet package, which analyzes high-dimensional (time series) data obtained from high-throughput functional genomics assays. GeneNet includes a computational algorithm that decides which edges are to be included in the final network, depending on the relative values of the pairwise partial correlations. Hence the output is a graph, where each gene corresponds to a node; the edges included in the graph portray direct dependencies between them. GeneNet estimates GGMs from small sample high-dimensional data relying on analytic shrinkage estimation of covariance and (partial) correlation matrices, and on model selection using (local) false discovery rate multiple testing. The approach is an exploratory method that may help to identify interesting genes or clusters of genes that are functionally related or coregulated, and may not necessarily yield the precise network of mechanistic interactions.

A summary on various undirected and directed network inference methods is reported in Table 6.2.

6.4 Inference assessment designs

The network inference methods are assessed in terms of their ability to infer networks as close to the real networks. Similar types of assessment are performed in other fields of bioinformatics too. For example, in the prediction of protein structures, when a researcher determines the spatial structure of a protein, it is compared to the real structure revealed by *in vitro* experiments. However, when it comes to gene regulatory networks, there are no gold standard experiments available to aid in determining the actual network structure that would serve as the gold standard for assessing an inference method using real networks. Therefore almost all assessments are done with the help of synthetic datasets, for which the gold standard network is already known [44] [50] [52].

Table 6.2 Synopsis of GRN inference methods.

Approach	Algorithm	Inferred network type	Package	Platform	Assessment based on (type of data)	Availability
Mutual Information	ARACNE (2006)	Undirected	MINET	R	Synthetic (Random network) & Real (Human B Cells)	https://www.bioconductor.org/packages/minet/
	CLR (2007)	Undirected	MINET	R	Real (E.coli)	https://www.bioconductor.org/packages/minet/
	MRNET (2007)	Undirected	MINET	R	Synthetic (sRogers & SynTReN)	https://www.bioconductor.org/packages/minet/
	C3NET (2010)	Undirected	C3NET	R	Synthetic (E.coli) & Real (E.coli)	https://cran.r-project.org/package=c3net
Expression Pattern Similarity	GeCON (2013)	Undirected	GeCON	Java	Synthetic(DREAM4) & Real (Yeast, Human, Rat, Mouse, Rice)	https://sites.google.com/site/swarupnehu/publications/resources
Multivariate Information Theory	PIDC (2017)	Undirected	PIDC	Julia	Synthetic (GNW- E. coli and Yeast)	https://github.com/Tchanders/network_inference_tutorials
Statistical Measures based Network	WGCNA (2008)	Undirected	WGCNA	R	-	http://www.genetics.ucla.edu/labs/horvath/CoexpressionNetwork/Rpackages/WGCNA
	GGM (2006)	Undirected	GeneNet	R	Synthetic (statically simulated) & Real (Breast Cancer)	http://strimmerlab.org/software/genenet/
Bayesian Network	LBN (2016)	Directed	-	-	Synthetic (DREAM3) & Real (E. coli)	-
Granger Causality	SWING (2018)	Directed	SWING	Python	Synthetic (E. coli and S. cerevisiae from GeneNetWeaver) & Real (E. coli and S. cerevisiae from RegulonDb and DREAM5 Yeast gold standards (Network4))	https://github.com/bagherilab/SWING
	PGC (2017)	Directed	prophetic-Genie3	R	Synthetic (HPN DREAM challenge data)	https://github.com/decarlin/prophetic-granger-causality
Regression-based Networks	TIGRESS (2012)	Directed	TIGRESS	MATLAB®	Synthetic & Real (E. coli and S. cerevisiae)	http://cbio.ensmp.fr/tigress
	SINCERITIRES (2017)	Directed	SINCERITIRES	MATLAB, R	Synthetic (E.coli and S.cerevisiae) Real (THP-1 human myeloid leukemia cell and T2EC chicken erythrocytic cells)	http://www.cabsel.ethz.ch/tools/sincerities.html https://github.com/CABSEL/SINCERITIES
Boolean Network & Genetic Algorithms	GABNI (2018)	Directed	-	-	Synthetic (Random Networks) & Real (DREAM Challenge (E.coli and Yeast) and Budding Yeast cell cycle network)	-
Boolean Network & Mutual Information	MIBNI (2017)	Directed	-	-	Synthetic (Random Networks) & Real (E.coli and fission yeast cell cycle network)	-
Dynamic Bayesian Network & ODE Models	DELDBN (2011)	Directed	-	R	Synthetic (Yeast IRMA network and human HeLa cell time series dataset)	https://academic.oup.com/bioinformatics/article/27/19/2686/231110#supplementary-data
Random Forest	GENIE3 (2010)	Directed	GENIE3	R, MatLab	Synthetic (DREAM4) & Real (E.coli)	https://bioconductor.org/packages/devel/bioc/html/GENIE3.html
Decision Tree & Boolean Network	yJUMP3 (2019)	Directed	Jump3	Matlab	DREAM4, IRMA	https://github.com/vahuynh/Jump3

6.4.1 Assessment against gold standard

Most of the assessments of GRN inference methods are performed using the networks developed during dialogue on reverse engineering assessment and methods (DREAM) challenges [64], which is a challenge that invite researchers from all over the world to propose solutions to questions in biology and medicine. Every year the organizers provide datasets to the participant community; the members of the respective communities are then expected to run their algorithms on those datasets and come up with a candidate network for each of the dataset provided. The organizers collect these networks and analyze them.

Network analysis is performed by comparing the candidate network submitted with the *a priori* determined network, also referred to as *reference*, or *gold standard* network. The measure of distance that indicates how close or far the candidate network is from the gold network is determined using the *confusion matrix*, which is a matrix that specifies the values of true positives (TP), true negatives (TN), false positives (FP), and false negatives (FN). The prediction accuracy against gold stand actually follows the same procedure as commonly used in any machine learning for assessing the accuracy of a classifiers.Thus assessment involves counting the following parameters:

- TP: the number of links correctly predicted by the algorithm.
- FP: the number of incorrectly predicted links.
- FN: the number of true links missed in the inferred network.
- TN: the number of correctly identified nonlinks.

Performance of an algorithm may then be summarized by calculating recall or true positive rate (TPR), false positive rate (FPR), precision or positive predictive value (PPV) or accuracy (Acc). They can be calculate as follows:

$$Recall = TP/(TP + FN) \tag{6.5a}$$

$$FPR = FP/(FP + TN) \tag{6.5b}$$

$$Precision = TP/(TP + FP) \tag{6.5c}$$

$$Acc = (TP + TN)/(TP + TN + FP + FN). \tag{6.5d}$$

The candidate inference method is usually assessed over a range of thresholds and, finally, the overall performance is summarized by using following parameters:

- ROC: The receiver operating characteristic (ROC) curve, which plots the 1-FPR (equivalent to the specificity) versus the TPR for all thresholds.
- PRC: the precision-recall (PR) curve, which plots the TPR against the PPV for all thresholds.

- F-measure: The F measure is a harmonic mean of precision, and recall can be calculated as $F_1 = (2 \cdot Recall \cdot Precision)/(Recall + Precision)$. An alternative and generic F measure is F_β to give β times as much importance to recall as precision. It can be calculated as follows:

$$F_\beta = (1 + \beta^2) \cdot \frac{precision \cdot recall}{(\beta^2 \cdot precision) + recall}. \qquad (6.6)$$

ROC curve is a plot of a function, where, on the x-axis, the false positive rate (FPR) and, on the y-axis, the true positive rate (TPR) are applied. The ROC curve represents the ratio between sensitivity and (1- specificity). When the ROC curve is above the line $y = x$, the classification is better.

To facilitate comparison of inference capabilities, instead of the ROC curve, the area under ROC curve (AUROC) can be used (sample AUROC or AUC is given in Fig. 6.3). The AUROC is the area covered by the ROC curve with the x-axis. The statistical meaning of the AUROC corresponds to the probability that the classifier will rank a randomly selected positive instance higher than a selected negative instance.

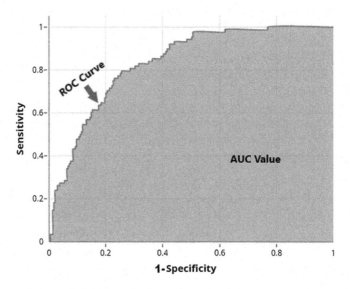

Figure 6.3. A sample ROC curve and the area under the curve shown as a shaded region.

Although the DREAM challenges are considered the *de facto* standard for the assessment of inference algorithms, few works point out its drawbacks. The DREAM challenges overlook the assumption [62] that the network inference problem is an undetermined problem, because experimental data alone is not sufficient to reconstruct a network completely. Simply put, the results obtained from a single microarray experiment do not cover all the biological molecules. For example, some microarray platforms do not cover all the transcriptomes; some transcripts could be missing; not to forget, the impact of the number and types of gene perturbation experiments, the noise that may occur in the data, and also the errors that may creep in during the experiments. All of these make the network inference problem an undetermined problem. Furthermore, investigatation [62] reports that there are some gene interactions that cannot be inferred and instances, where more than one network can agree with the data.

To cope with these issues, a novel assessment procedure has been proposed that checks the inferability of gene regulatory interactions by means of a confusion matrix that specifies the inferability of the network. In other words, the possibility of the network to be determined from data. The causal information extracted from experiments is used to analyze the inferability of GRNs. Herein, data from the DREAM 4 *in silico* network inference challenge is used to score the assessment and to show the concept of inferability from experiments. A new performance score is thus introduced based on the redefinition of the confusion matrix that also considers the noninferable gene regulations.

6.4.2 Assessment in absence of true networks

In addition to the existing challenges faced by a system biologist, the selection of a suitable inference method pose another hurdle, because of the limited availability of real networks of living organisms to aid in the validation of various inference methods. Most of the researchers use already known interaction that have been validated by previous researches to validate the networks obtained from real data. Some also use experimentally validated regulators that can be collected from publicly available databases, such as RegulonDB [25], to assess the performance of inference methods [31]. RegulonDB 9.0 includes 184 experimentally determined transcription factors (TFs) and 120 computationally predicted TF for *E. coli* only. Pathways are also useful in this regard, because they provide a way of linking the functionality of group of genes to specific biological processes. In this regard, well-established methodologies, such as gene set enrichment analysis (GSEA) [65] can help in differentiating pathways as functional units from experimental populations. Kyoto encyclopedia of genes and genomes (KEGG) provide manually curated pathways based on expert knowledge and existing literature, and can be used as an alternative measure for validation [6]. An alternative measure based on gene ontology (GO) against functional, biological enrichment of a group of genes derived from inferred network modules [56] can also be used in evaluating the biological significance of an inference method. Herein, the more a method is able to generate biologically significant modules (in terms of p value), the more they are considered relevant.

6.4.3 Experimenting with few inference methods

To demonstrate how good inference methods are and how effective they are in inferring network close to any real network, we consider four methods for experiments: ARACNE, CLR, MRNET, and GENE3. We use E.coli networks and the corresponding synthetic expression profiles, reverse engineered by GeneNetWeaver (GNW) [61]. The inferred networks are compared with the gold standard and the overall performance reported in terms of AUROC (Fig. 6.4) and AUPR (Fig. 6.5) values. To assess how good is the inference quality in producing networks close to real world networks, we compute the characteristics of the inferred networks for three parameters, such as cumulative degree distribution (Fig. 6.6), average path length (Fig. 6.7), and clustering coefficient (Fig. 6.8). From the figure we may conclude that the degree distribution of the networks inferred by candidate methods show power-law curve, which indicates that they are producing scale-free networks.

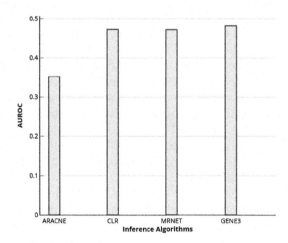

Figure 6.4. Performance of four candidate inference methods on E.Coli expression network based on AUROC.

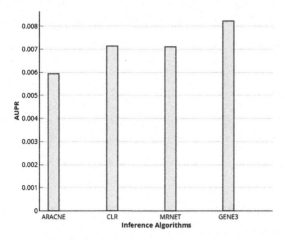

Figure 6.5. Area under precision-recall for different coexpression network methods on DREAM data (E. Coli).

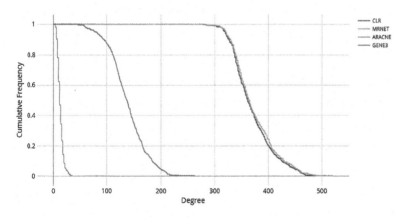

Figure 6.6. Degree distribution of nodes in the inferred networks shows power-law properties.

6.5 Post inference network analysis

Any inference methods usually predict a graph based on the association used. However, the entire network may not be very meaningful in playing a biological role. Rather, a compact subnetwork may play cohesive function in certain disease or biological conditions. The extraction of such compact subnetwork, termed as module detection, is thus a challenging work, involving the isolation of more focused subsets of genes.

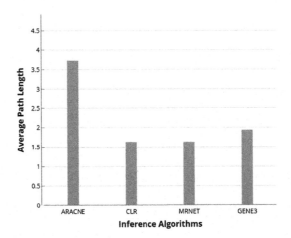

Figure 6.7. Average path length of the inferred networks.

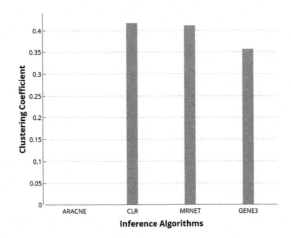

Figure 6.8. Clustering coefficient of the networks inferred by the candidate methods.

6.5.1 Network module detection

A set of correlated and coexpressed genes, often referred as a functional module, play a synergistic role during any disease or any biological activity. Genes participating in a common module may cause clinically similar diseases and shares the common genetic origin of their associated disease phenotypes. Identifying such modules may be helpful in system-level understanding of biological and cellular processes or the pathophysiologic basis of associated diseases. Formally, we can define a network module as follows:

Definition 6.5.1 (Network module). Given a network $\mathcal{G}$, a network module $\mathcal{M}_i = \{\mathcal{V}', \mathcal{E}'\}$ is a densely connected subgraph of $\mathcal{G}$ ($\mathcal{M}_i \subseteq \mathcal{G}$), where interconnectivity of $\mathcal{V}'$ with respect to $\mathcal{E}' \subseteq \mathcal{E}$ is higher in comparison to the rest of $\mathcal{V}$, i.e., $\mathcal{V} - \mathcal{V}'$.

The first step in this analysis is the building of (weighted or unweighted) graph starting from experimental data. Next, a network module or community detection method is applied. Community discovery algorithm may be categorized using different parameters [24], e.g., on the nature of discovered modules (overlapping or not), on their structure (densely connected subgraph, graphlet-based). Here, we do not propose any other classification, and we selected some state-of-the-art algorithms, and we categorized them into two broad classes: (i) algorithms developed specifically for gene expression analysis, and (ii) algorithm for network analysis that may be used for such networks.

WGCNA [37] is a popular method to detect modules from gene networks. It receives the coexpression network as input representing correlations, and it applies a soft thresholding to remove the possibility of non-relevant edges under the hypothesis that communities are made of relevant edges. After the thresholding, it employs a fuzzy approach to extract (possibly overlapping) modules without any hypothesis on the internal structure. The method proposed in [58], builds a correlation network first using an ad-hoc method, and then it employs a spectral clustering to mine the obtained network. Therefore, it receives as input raw gene expression data, and it can find clusters without imposing any constraint on the structure. As in the case of the previous method, the FUMET (fuzzy network module extraction technique) algorithm [42] proposes a novel method for the construction of coexpression network and a network module extraction technique based on fuzzy set theoretic approach. It can handle both positive and negative correlations among genes. Module miner [41] is similar to FUMET in the building of correlation network, and it employs a different module extraction approach. The molecular complex detection (MCODE) algorithm [4] is initially developed to analyze protein interaction networks. It searches for densest connected subgraphs and does not admit overlapping among discovered modules. The Markov CLustering (MCL) [21] is an well-known algorithm used to find clusters on graphs. It is robust to noise and graph alterations. Brohee and Van Helden demonstrated in an extensive comparison [11] that MCL outperforms other clustering algorithms in different conditions and using suboptimal parameters [15]. The rationale behind MCL is quite simple. A possible way to define a module within a network is a collection of nodes that are more connected to each other than to the others. It follows that a random walk starting in any of these nodes is more likely to stay within the cluster rather than to travel between clusters. By simulating many random walks starting from the various nodes, it is possible to identify flows of random walks that tend to gather in specific regions of the graph (the modules). MCL [19,71] is an iterative algorithm that simulates random walks using Markov chains. Table 6.3 lists few of the network based algorithms.

Table 6.3 Network based module finding algorithms.

Algorithm	Model	Availability	Time complexity	Nature of modules
Spectral based [58]	Spectral clustering algorithm	-	-	Exclusive
WGCNA [37]	Hierarchical clustering	http://www.genetics.ucla.edu/labs/horvath/ CoexpressionNetwork/Rpackages/WGCNA	$O(n^2)$, n is no. of genes	Nonexclusive
MCL [19,71]	Markov chain	http://www.cytoscape.org/	$O(n^3)$, n is no. of nodes of the input graph	-
MCODE [4]	Density based	ftp://ftp.mshri.on.ca/pub/BIND/Tools/MCODE, http://www.cytoscape.org	$O(nmh^3)$, n is no. of vertices, m is no. of edges, h is vertex size of the avg vertex neighborhood in the input graph G.	Nonexclusive
Module miner [41]	Spanning tree	-	$O(n^2)$, n is no. of genes	Nonexclusive
FUMET [42]	Fuzzy	-	$O(n^2)$, n is no. of genes	Nonexclusive

A detailed discussion on various module detection techniques and their comparative performance assessments can be found in [33].

6.5.2 Ranking key diseased genes using network analysis

To study the causes of complex diseases, researchers focus on detecting subnetwork of functionally interrelated genes forming a functional module. However, not all the genes within a module play key roles in disease formation. Rather, a very few genes are the pivotal genes. The latter are called *marker genes*. They are responsible for disrupting the normal cellular functionalities, causing diseases. They are often identified as transcription factor (TF) genes. TF binds with the promoter region of target genes and lead to abnormal expression of the genes. Identifying such key genes responsible for the formation of disease networks may help in designing disease-specific drugs. A number of prioritization schemes have been proposed in different literature. Majority of them adopt centrality analysis of the disease subnetworks. It has been observed that the outcome of such biomarker ranking or prioritization scheme is sensitive towards the input network.

Detection of marker genes responsible for a genetic disease is a difficult task. Many researchers have dedicated their work in detecting such genes using various ranking techniques. Cluvian [43] identifies key genes that are possibly responsible for Alzheimer's disease by analyzing modules derived from Alzheimer's disease (AD) coexpression networks. The networks first extract AD submodules and rank them based on AD pathway enrichment scores. Top ranked modules are further analyzed topologically to identify central or hub genes, which are the potential key genes responsible for AD. In [39,48], they devised a ranking scheme using varied correlation measurements for the improvement of microarray and RNA-seq-based global and targeted coexpression networks. In addition to ordering genes based on fold change across the data, they also consider all three cell type-associated measures. In another attempt, authors considered a gene as a marker gene when genes are differentially expressed during some conditions or during protein interaction [74]. HyDRA (hybrid distance-score rank aggregation) [35], applies score-based and combinatorial aggregation techniques. It integrates a top-versus-bottom (TvB) weighting feature into the hybrid schemes. Using this scheme, it considers only top candidate genes. Biomarker ensemble ranking framework (BERF) [16] is developed for the detection of genes responsible for depression. This method employs two ranking models. It considers genes, which are already known marker and nonmarker genes. For a generation of ranking results, it uses an ensemble technique. HetRank [17] is a technique used for ranking gene on interaction network data. The algorithm focuses on two folds; the first fold concentrates on showing that genes triggering a disease are usually interconnected in PPI networks; whereas in the second fold, the method concentrates on genes expressing with varied pathogenic variations and their neighboring genes are marker genes.

To effectively analyze a network, it is convenient to have an interactive view of the network for the ease of the better analysis. A number of visualization and network analysis tools are freely available, which may facilitate effective biological network research.

6.6 Network visualization and analysis tools

A number of *in silico* visualization and analysis tools are available to provide user-friendly environment for the system biologist. However, some of the tools are even equally applicable in other type of networks too. They are either online web-based or stand-alone desktop versions, which differ from one another in their way of generating and presenting the networks. Most of the effort is the outcome of various researches in the area of GRN inference methods or techniques. They are normally limited to any single inference method. Some of the tools also provide benchmarking and synthetic data-generation facilities. Other than inference of GRN, visualization of networks are also considered as an integral component of the tools. Below we present a comprehensive study on various tools and review their features in order to help the biologists select appropriate tools that may suit their own requirements.

GeneNetWeaver   GeneNetWeaver (GNW) [61] is a tool developed in Java for *in silico* benchmarking and performance evaluation of network inference methods. Benchmarking involves generating gene network structures; generating simulated data from these networks using adequate dynamical models. In GNW, subnetwork extraction starts with the extraction of modules, which are groups of genes that are highly connected in a random network from a given global interaction network. This tool is able to perform a network motif analysis from a set of network predictions and their corresponding benchmark networks. The accuracy of network inference can be assessed using standard metrics, such as precision-recall (PR) and receiver operating characteristic (ROC) curves.

Cytoscape   Cytoscape [63] is a desktop complex network analysis and visualization tool in life sciences. Additional features are available as plugins. For example, Cytoscape can visualize molecular interaction networks, and integrate with gene expression profiles and other state data. Cytoscape has three versions: Cytoscape 2.x, Cytoscape 3.x, and Cytoscape.js. Cytoscape.js is a successor of CytoscapeWeb. This tool is most suitable for large-scale network analysis since it can handle thousands of nodes and edges and still run smoothly. Cytoscape supports directed, undirected, and weighted graphs and comes along with powerful visual styles, thereby allowing users to change the properties of nodes or edges. Plenty of elegant layout algorithms, including cyclic and spring-embedded layouts are available for visualization. Expression data can be mapped as node color, label, border thickness, or border color.

GENeVis   GENeVis [5], is a Java desktop application that allows simultaneously visualizing gene regulatory networks and gene expression time series data. While running GENeVis for the first time, internet connection is required to load information from the KEGG database server. All data files are in plain text, and are represented as tables with tab-separated columns.

LegumeGRN   LegumeGRN [69] is a GRN analysis tool for legume species. The web-server of this tool is preloaded with gene expression data for medicago, lotus and soybean. LegumeGRN is equipped with four types of GRN prediction algorithms, i.e., Graphical Gaussian Models (GGMs), Context Likelihood of Relatedness (CLR) [22], TIGRESS [30], and GENIE3 [32]. Users are able to select which experiments, genes, and GRN algorithms they wish to perform for GRN analysis. Users can visualize results of the experiment based on selected algorithm(s).

STARNET2   STARNET2 [34] is a web-based tool for accelerating discovery of gene regulatory networks using microarray data. STARNET2 facilitates discovery of putative gene regulatory networks in a variety of species (*Homo sapiens, Rattus norvegicus, Mus musculus, Gallus gallus, Danio rerio, Drosophila, C. elegans, S. cerevisiae, Arabidopsis and Oryza Sativa*) by graphing networks of genes that are closely coexpressed across a large heterogeneous set of preselected microarray experiments. The precompiled results are stored in a MySQL database. The focus of this tool is on the user's queries of the database centered on a gene of interest. The user should know the gene symbol or the Entrez gene ID, which he or she intended to search for. The output includes graphs of correlation networks, graphs of known interactions involving genes and gene products that are present in the correlation networks, and initial statistical analyses.

NetBioV   NetBioV [67] (network biology visualization) is an R package that allows visualization of large network data in biology and medicine. This tool is available freely for academic purposes from Bioconductor. The purpose of NetBioV is to enable an organized and reproducible visualization of networks by emphasizing or highlighting specific structural properties that are of biological relevance.

FastMEDUSA   Fast MEDUSA [9] is a parallel program to infer gene regulatory networks from gene expression and promoter sequences which is implemented in C++, utilizing the message passing interface (MPI) implementation, MPICH2, for interprocess communication and GotoBLAS2 library for performing matrix multiplication. It can be installed on Mac OS X and Linux platforms.

6.7 Summary

Gene network inference is still in its infancy due to lack of proper validated true networks. The regulation process, in reality, is a complex multi-step process occurring inside a cell. Inference methods developed based on mathematical or statistical tools alone are not adequate to capture the natural phenomena of regulation. As a result, regulatory network requires multisource experimental data for better outcomes. One more important factor is the computational resource constraints. Usually, the number of genes is large. Almost all the inference methods perform pairwise association (or causality) computation. Hence highly expensive while handling genome-scale networks with large number of genes, in the scale of 20 K to 40 K in any standard computers. Parallel inference is an effective alternative to address the bottleneck. Future research in this domain may largely concentrate on genome-scale network inference both from multi-cell and single-cell expression data. Causal edge inference is another issue that needs to be addressed effectively in the near future.

Acknowledgment

Authors thank Prof. Dhruba K Bhattacharyya, Mr. Graciously Kharumnuid, Dr. Monica Jha, Ms Softya Sebastian, and Sk. Atahar Ali for their valuable suggestions and contributions in the writing of this chapter.

References

1. Shiek S.S.J. Ahmed, Winkins Santosh, Suresh Kumar, Hema T. Thanka Christlet, Metabolic profiling of Parkinson's disease: evidence of biomarker from gene expression analysis and rapid neural network detection, Journal of Biomedical Science 16 (1) (2009) 63.
2. Syed Sazzad Ahmed, Swarup Roy, Jugal K. Kalita, Assessing the effectiveness of causality inference methods for gene regulatory networks, in: IEEE/ACM Transactions on Computational Biology and Bioinformatics, 2018.
3. Gökmen Altay, Frank Emmert-Streib, Inferring the conservative causal core of gene regulatory networks, BMC Systems Biology 4 (1) (2010) 132.
4. Gary D. Bader, Christopher W.V. Hogue, An automated method for finding molecular complexes in large protein interaction networks, BMC Bioinformatics 4 (1) (2003) 2.

5. C.A. Baker, M.S.T. Carpendale, Przemyslaw Prusinkiewicz, Michael G. Surette, Genevis: visualization tools for genetic regulatory network dynamics, in: Proceedings of the Conference on Visualization'02, IEEE Computer Society, 2002, pp. 243–250.

6. Nikolay Balov, A categorical network approach for discovering differentially expressed regulations in cancer, BMC Medical Genomics 6 (3) (2013) S1.

7. Shohag Barman, Yung-Keun Kwon, A novel mutual information-based Boolean network inference method from time-series gene expression data, PLoS ONE 12 (2) (2017) e0171097.

8. Shohag Barman, Yung-Keun Kwon, A Boolean network inference from time-series gene expression data using a genetic algorithm, Bioinformatics 34 (17) (2018) i927–i933.

9. Serdar Bozdag, Aiguo Li, Stefan Wuchty, Howard A. Fine, Fastmedusa: a parallelized tool to infer gene regulatory networks, Bioinformatics 26 (14) (2010) 1792–1793.

10. Paul Brazhnik, Alberto de la Fuente, Pedro Mendes, Gene networks: how to put the function in genomics, Trends in Biotechnology 20 (11) (2002) 467–472.

11. Sylvain Brohee, Jacques Van Helden, Evaluation of clustering algorithms for protein-protein interaction networks, BMC Bioinformatics 7 (1) (2006) 488.

12. Atul J. Butte, Isaac S. Kohane, Relevance networks: a first step toward finding genetic regulatory networks within microarray data, in: The Analysis of Gene Expression Data, Springer, 2003, pp. 428–446.

13. Daniel E. Carlin, Evan O. Paull, Kiley Graim, Christopher K. Wong, Adrian Bivol, Peter Ryabinin, Kyle Ellrott, Artem Sokolov, Joshua M. Stuart, Prophetic granger causality to infer gene regulatory networks, PLoS ONE 12 (12) (2017) e0170340.

14. Thalia E. Chan, Michael P.H. Stumpf, Ann C. Babtie, Gene regulatory network inference from single-cell data using multivariate information measures, Cell Systems 5 (3) (2017) 251–267.

15. Young-Rae Cho, Marco Mina, Yanxin Lu, Nayoung Kwon, Pietro H. Guzzi, M-finder: uncovering functionally associated proteins from interactome data integrated with go annotations, Proteome Science 11 (1) (2013) S3.

16. Abu Sayed Chowdhury, Md Monjur Alam, Yanqing Zhang, A biomarker ensemble ranking framework for prioritizing depression candidate genes, in: Computational Intelligence in Bioinformatics and Computational Biology (CIBCB), 2015 IEEE Conference on, IEEE, 2015, pp. 1–6.

17. Nick Dand, Reiner Schulz, Michael E. Weale, Laura Southgate, Rebecca J. Oakey, Michael A. Simpson, Thomas Schlitt, Network-informed gene ranking tackles genetic heterogeneity in exome-sequencing studies of monogenic disease, Human Mutation 36 (12) (2015) 1135–1144.

18. S. Das, D. Caragea, S.M. Welch, W.H. Hsu, Handbook of Research on Computational Methodologies in Gene Regulatory Networks, Medical Information Science Reference, 2010.

19. S. Van Dongen, Graph clustering by flow simulation, PhD thesis, Standardization and Knowledge Transfer, 2000.

20. Bradley Efron, Trevor Hastie, Iain Johnstone, Robert Tibshirani, et al., Least angle regression, The Annals of Statistics 32 (2) (2004) 407–499.

21. Anton J. Enright, Stijn Van Dongen, Christos A. Ouzounis, An efficient algorithm for large-scale detection of protein families, Nucleic Acids Research 30 (7) (2002) 1575–1584.

22. Jeremiah J. Faith, Boris Hayete, Joshua T. Thaden, Ilaria Mogno, Jamey Wierzbowski, Guillaume Cottarel, Simon Kasif, James J. Collins, Timothy S. Gardner, Large-scale mapping and validation of escherichia coli

transcriptional regulation from a compendium of expression profiles, PLoS Biology 5 (1) (2007) e8.

23. Justin D. Finkle, Jia J. Wu, Neda Bagheri, Windowed granger causal inference strategy improves discovery of gene regulatory networks, Proceedings of the National Academy of Sciences 115 (9) (2018) 2252–2257.

24. Santo Fortunato, Community detection in graphs, Physics Reports 486 (3) (2010) 75–174.

25. Socorro Gama-Castro, Verónica Jiménez-Jacinto, Martin Peralta-Gil, Alberto Santos-Zavaleta, Mónica I. Peñaloza-Spinola, Bruno Contreras-Moreira, Juan Segura-Salazar, Luis Muniz-Rascado, Irma Martinez-Flores, Heladia Salgado, et al., Regulondb (version 6.0): gene regulation model of escherichia coli k-12 beyond transcription, active (experimental) annotated promoters and textpresso navigation, Nucleic Acids Research 36 (suppl 1) (2008) D120–D124.

26. A.P. Gasch, M.B. Eisen, et al., Exploring the conditional coregulation of yeast gene expression through fuzzy k-means clustering, Genome Biology 3 (11) (2002) 1–22.

27. Clive W.J. Granger, Investigating causal relations by econometric models and cross-spectral methods, Econometrica: Journal of the Econometric Society (1969) 424–438.

28. R.P. Grant, Computational Genomics: Theory and Application, Horizon Bioscience, 2004.

29. Yan Guo, Chung-I. Li, Fei Ye, Yu Shyr, Evaluation of read count based rnaseq analysis methods, BMC Genomics 14 (8) (2013) S2.

30. Anne-Claire Haury, Fantine Mordelet, Paola Vera-Licona, Jean-Philippe Vert, Tigress: trustful inference of gene regulation using stability selection, BMC Systems Biology 6 (1) (2012) 145.

31. Vân Anh Huynh-Thu, Alexandre Irrthum, Louis Wehenkel, Yvan Saeys, Pierre Geurts, Inferring gene regulatory networks from expression data using tree-based methods, 2011.

32. Alexandre Irrthum, Louis Wehenkel, Pierre Geurts, et al., Inferring regulatory networks from expression data using tree-based methods, PLoS ONE 5 (9) (2010) e12776.

33. Monica Jha, Pietro H. Guzzi, Swarup Roy, Qualitative assessment of functional module detectors on microarray and rnaseq data, Network Modeling Analysis in Health Informatics and Bioinformatics 8 (1) (2019) 1.

34. Daniel Jupiter, Hailin Chen, Vincent VanBuren, Starnet 2: a web-based tool for accelerating discovery of gene regulatory networks using microarray co-expression data, BMC Bioinformatics 10 (1) (2009) 332.

35. Minji Kim, Farzad Farnoud, Olgica Milenkovic, Hydra: gene prioritization via hybrid distance-score rank aggregation, Bioinformatics 31 (7) (2014) 1034–1043.

36. A. Kommadath, M.F.W. te Pas, M.A. Smits, Gene coexpression network analysis identifies genes and biological processes shared among anterior pituitary and brain areas that affect estrous behavior in dairy cows, Journal of Dairy Science 96 (4) (2013) 2583–2595.

37. Peter Langfelder, Steve Horvath, Wgcna: an r package for weighted correlation network analysis, BMC Bioinformatics 9 (1) (2008) 559.

38. Zheng Li, Ping Li, Arun Krishnan, Jingdong Liu, Large-scale dynamic gene regulatory network inference combining differential equation models with local dynamic Bayesian network analysis, Bioinformatics 27 (19) (2011) 2686–2691.

39. Franziska Liesecke, Dimitri Daudu, Rodolphe Dugé de Bernonville, Sébastien Besseau, Marc Clastre, Vincent Courdavault, Johan-Owen

De Craene, Joël Crèche, Nathalie Giglioli-Guivarc'h, Gaëlle Glévarec, et al., Ranking genome-wide correlation measurements improves microarray and rna-seq based global and targeted co-expression networks, Scientific Reports 8 (1) (2018) 10885.

40. Fei Liu, Shao-Wu Zhang, Wei-Feng Guo, Ze-Gang Wei, Luonan Chen, Inference of gene regulatory network based on local Bayesian networks, PLoS Computational Biology 12 (8) (2016) e1005024.

41. Priyakshi Mahanta, Hasin A. Ahmed, Dhruba K. Bhattacharyya, Jugal K. Kalita, An effective method for network module extraction from microarray data, BMC Bioinformatics 13 (13) (2012) S4.

42. Priyakshi Mahanta, Hasin Afzal Ahmed, Dhruba Kumar Bhattacharyya, Ashish Ghosh, Fumet: a fuzzy network module extraction technique for gene expression data, Journal of Biosciences 39 (3) (2014) 351–364.

43. Hazel Nicolette Manners, Swarup Roy, Jugal K. Kalita, Intrinsic-overlapping co-expression module detection with application to Alzheimer's disease, Computational Biology and Chemistry 77 (2018) 373–389.

44. Daniel Marbach, Thomas Schaffter, Claudio Mattiussi, Dario Floreano, Generating realistic in silico gene networks for performance assessment of reverse engineering methods, Journal of Computational Biology 16 (2) (2009) 229–239.

45. Adam A. Margolin, Ilya Nemenman, Katia Basso, Chris Wiggins, Gustavo Stolovitzky, Riccardo Dalla Favera, Andrea Califano, Aracne: an algorithm for the reconstruction of gene regulatory networks in a mammalian cellular context, BMC Bioinformatics 7 (2006) S7.

46. Andrea Massaia, Patricia Chaves, Sara Samari, Ricardo Júdice Miragaia, Kerstin Meyer, Sarah Amalia Teichmann, Michela Noseda, Single cell gene expression to understand the dynamic architecture of the heart, Frontiers in Cardiovascular Medicine 5 (2018) 167.

47. Hirotaka Matsumoto, Hisanori Kiryu, Chikara Furusawa, Minoru S.H. Ko, Shigeru B.H. Ko, Norio Gouda, Tetsutaro Hayashi, Itoshi Nikaido, Scode: an efficient regulatory network inference algorithm from single-cell rna-seq during differentiation, Bioinformatics 33 (15) (2017) 2314–2321.

48. Andrew T. McKenzie, Minghui Wang, Mads E. Hauberg, John F. Fullard, Alexey Kozlenkov, Alexandra Keenan, Yasmin L. Hurd, Stella Dracheva, Patrizia Casaccia, Panos Roussos, et al., Brain cell type specific gene expression and co-expression network architectures, Scientific Reports 8 (2018).

49. Nicolai Meinshausen, Peter Bühlmann, Stability selection, Journal of the Royal Statistical Society: Series B (Statistical Methodology) 72 (4) (2010) 417–473.

50. Pedro Mendes, Wei Sha, Keying Ye, Artificial gene networks for objective comparison of analysis algorithms, Bioinformatics 19 (suppl 2) (2003) ii122–ii129.

51. Patrick E. Meyer, Kevin Kontos, Frederic Lafitte, Gianluca Bontempi, Information-theoretic inference of large transcriptional regulatory networks, EURASIP Journal on Bioinformatics and Systems Biology 2007 (2007) 8.

52. John Moult, Krzysztof Fidelis, Andriy Kryshtafovych, Burkhard Rost, Tim Hubbard, Anna Tramontano, Critical assessment of methods of protein structure prediction—round vii, Proteins: Structure, Function, and Bioinformatics 69 (S8) (2007) 3–9.

53. Nan Papili Gao, S.M. Minhaz Ud-Dean, Olivier Gandrillon, Rudiyanto Gunawan, Sincerities: inferring gene regulatory networks from time-stamped single cell transcriptional expression profiles, Bioinformatics 34 (2) (2017) 258–266.

54. Hugues Richard, Marcel H. Schulz, Marc Sultan, Asja Nurnberger, Sabine Schrinner, Daniela Balzereit, Emilie Dagand, Axel Rasche, Hans Lehrach, Martin Vingron, et al., Prediction of alternative isoforms from exon expression levels in rna-seq experiments, Nucleic Acids Research 38 (10) (2010) e112.

55. Swarup Roy, Dhruba K. Bhattacharyya, Jugal K. Kalita, Cobi: pattern based co-regulated biclustering of gene expression data, Pattern Recognition Letters 34 (14) (2013) 1669–1678.

56. Swarup Roy, Dhruba K. Bhattacharyya, Jugal K. Kalita, Reconstruction of gene co-expression network from microarray data using local expression patterns, BMC Bioinformatics 15 (7) (2014) S10.

57. Swarup Roy, Pooja Sharma, Keshab Nath, Dhruba K. Bhattacharyya, Jugal K. Kalita, Pre-processing: a data preparation step, in: Shoba Ranganathan, Michael Gribskov, Kenta Nakai, Christian Schönbach (Eds.), Encyclopedia of Bioinformatics and Computational Biology, Academic Press, Oxford, 2019, pp. 463–471.

58. Jianhua Ruan, Weixiong Zhang, Identification and evaluation of functional modules in gene co-expression networks, in: Systems Biology and Computational Proteomics, Springer, 2007, pp. 57–76.

59. Guido Sanguinetti, et al., Tree-based learning of regulatory network topologies and dynamics with jump3, in: Gene Regulatory Networks, Springer, 2019, pp. 217–233.

60. Juliane Schäfer, Rainer Opgen-Rhein, Korbinian Strimmer, Reverse engineering genetic networks using the genenet package, Newsletter of the R Project 6 (5) (December 2006) 50–53.

61. Thomas Schaffter, Daniel Marbach, Dario Floreano, Genenetweaver: in silico benchmark generation and performance profiling of network inference methods, Bioinformatics 27 (16) (2011) 2263–2270.

62. Caroline Siegenthaler, Rudiyanto Gunawan, Assessment of network inference methods: how to cope with an underdetermined problem, PLoS ONE 9 (3) (2014) e90481.

63. Michael E. Smoot, Keiichiro Ono, Johannes Ruscheinski, Peng-Liang Wang, Trey Ideker, Cytoscape 2.8: new features for data integration and network visualization, Bioinformatics 27 (3) (2011) 431–432.

64. Gustavo Stolovitzky, D.O.N. Monroe, Andrea Califano, Dialogue on reverse-engineering assessment and methods: the dream of high-throughput pathway inference, Annals of the New York Academy of Sciences 1115 (1) (2007) 1–22.

65. Aravind Subramanian, Pablo Tamayo, Vamsi K. Mootha, Sayan Mukherjee, Benjamin L. Ebert, Michael A. Gillette, Amanda Paulovich, Scott L. Pomeroy, Todd R. Golub, Eric S. Lander, et al., Gene set enrichment analysis: a knowledge-based approach for interpreting genome-wide expression profiles, Proceedings of the National Academy of Sciences 102 (43) (2005) 15545–15550.

66. S. Tavazoie, J.D. Hughes, M.J. Campbell, R.J. Cho, G.M. Church, et al., Systematic determination of genetic network architecture, Nature Genetics 22 (1999) 281–285.

67. Shailesh Tripathi, Matthias Dehmer, Frank Emmert-Streib, Netbiov: an r package for visualizing large network data in biology and medicine, Bioinformatics 30 (19) (2014) 2834–2836.

68. Sipko van Dam, Urmo Vosa, Adriaan van der Graaf, Lude Franke, Joao Pedro de Magalhaes, Gene co-expression analysis for functional classification and gene–disease predictions, Briefings in Bioinformatics 19 (4) (2017) 575–592.

69. Mingyi Wang, Jerome Verdier, Vagner A. Benedito, Yuhong Tang, Jeremy D. Murray, Yinbing Ge, Jörg D. Becker, Helena Carvalho, Christian Rogers, Michael Udvardi, et al., Legumegrn: a gene regulatory network prediction server for functional and comparative studies, PLoS ONE 8 (7) (2013) e67434.

70. Zhong Wang, Mark Gerstein, Michael Snyder, RNA-Seq: a revolutionary tool for transcriptomics, Nature Reviews. Genetics 10 (1) (2009) 57.

71. Guanming Wu, Lincoln Stein, A network module-based method for identifying cancer prognostic signatures, Genome Biology 13 (12) (2012) R112.

72. Bin Zhang, Chris Gaiteri, Liviu-Gabriel Bodea, Zhi Wang, Joshua McElwee, Alexei A. Podtelezhnikov, Chunsheng Zhang, Tao Xie, Linh Tran, Radu Dobrin, et al., Integrated systems approach identifies genetic nodes and networks in late-onset Alzheimer's disease, Cell 153 (3) (2013) 707–720.

73. Bin Zhang, Steve Horvath, A general framework for weighted gene co-expression network analysis, Statistical Applications in Genetics and Molecular Biology 4 (1) (2005).

74. Jing Zhao, Ting-Hong Yang, Yongxu Huang, Petter Holme, Ranking candidate disease genes from gene expression and protein interaction: a Katz-centrality based approach, PLoS ONE 6 (9) (2011) e24306.

Protein interaction networks

Contents

One of the challenges in computational biology is the analysis of the so-called *interactome*, i.e., the whole network composed of proteins and their interactions. Proteins are elementary building blocks of complex cellular processes; therefore the identification of all the possible protein interactions, the elucidation of protein functions, and the role of a set of related interactions (*e.g., protein complexes and pathways*), do appear as essential steps in describing cellular biology on a molecular basis. In this chapter, we try to highlight different methods involved in the process of protein interaction generation, aligning various protein networks for finding evolutionarily conserved protein functionality, and small subnetwork findings.

7.1 Proteins and interaction graph

Protein molecules are the workhorses of the cell, performing and controlling almost all activities in an organism. Although some proteins may work alone, proteins usually collaborate with others to achieve their intended tasks. When proteins work together, the influences and interactions among them can be shown in terms of a graph. The human cell produces potentially 100,000 different proteins, where a gene may produce more than one pro-

Biological Network Analysis. https://doi.org/10.1016/B978-0-12-819350-1.00013-X

tein. The interactions among these proteins are responsible for many physiological activities in the body. Organisms vary in the number of their proteins and the number of interactions. Proteins and protein interactions of a large number of organisms are being determined, and are recorded in databases. According to one study [94], humans have ten times more protein interactions than the fruit fly, and 20 times more than the single-celled yeast. A simple but useful approach to viewing such a complex biological system is to represent it as a network of the interplay among the different molecules.

Thus, complex systems, such as protein-protein interactions (PPI), are usually studied computationally from a graph-theoretic perspective. Studies suggest that PPI networks (PIN) are conserved through evolution [89]. Highly connected proteins within a network are vital molecules and have been found to be more essential for survival than proteins with lower connectivity [43]. As a result, the interactions between protein pairs and the overall composition of the network are important for the overall functioning of an organism. Understanding conserved substructures through a comparative analysis of these networks can provide insights into a variety of biochemical processes. The ultimate goal of network alignment is to transfer knowledge of protein function from one species to another. Since sequence similarity metrics, such as BLAST bit scores [3], are not conclusive evidence of similar function, the purpose of aligning two PPI networks is to supplement sequence similarity with topological information so as to identify orthologs as accurately as possible.

The physical interaction between a pair of protein molecules takes place because of biochemical activities within a cell. The synthesis of protein may also be regulated by another protein. When proteins work together, the influences and interactions among them can be shown in terms of a graph [10]. A graph showing interactions among proteins in a single species is called a protein-protein interaction network (or map). In a PPI graph, the proteins are nodes, and molecular interactions between them are edges.

Definition 7.1.1 (Protein-protein-interaction network). A set of proteins $\mathcal{P}$ forms a PPI network $\mathcal{G} = (\mathcal{P}, \mathcal{I})$ by virtue of the physical interactions among the protein molecules in $\mathcal{P}$ due to various biochemical events. $\mathcal{I}$ represents the set of interactions between a pair of proteins P_i and $P_j \in \mathcal{P}$.

The set of interactions has been modeled by using graph theory, enabling the discovery of biologically relevant knowledge directly from these models. Main research fields have been (i) the individuation of a theoretical model of network, to which PINs belong; (ii) the analysis of single of PINs, and (iii) the comparison of PINs through alignment algorithms.

Despite the proposal of different models for the interactome, (such as the *random graph model* [9], the *scale-free model*, the *geometric model* [77], the STICKY model), there is still room for the research of a commonly accepted model for PINs (see for instance the recent works [58,61]). As noted in Cannataro et al., [12], each of these models suffers some problems, and, currently, every attempt of comparison has a drawback that is due to the missing knowledge of the real interaction map. Moreover, the structure seems to be related to species, e.g., interactome of viruses are significantly different from the other [26].

The analysis of properties of PINs and the extraction of interesting patterns of interactions has focused mainly in the *in silico* individuation of network modules encoding biological knowledge (e.g., protein complexes). Moreover, different methods for the automatic extraction of a set of related proteins performing the same biological function, i.e., pathways, have been introduced. Fig. 7.1 summarizes the flow of activities in interactomics research, from wet lab to knowledge extracted with *in silico* methods.

7.2 Methods for protein interaction generation

The interaction between a pair of protein molecules is measured using a variety of assays, such as immunoprecipitation, tandem affinity purification, and the yeast two-hybrid approach. Recently, these techniques have been scaled up to measure interactions on a genome-wide level [98]. High-throughput techniques have also been developed to systematically identify protein complexes using affinity purification techniques, followed by mass spectrometry (MS/MS) to sequence the proteins [21,35]. The process of building a protein interaction network starts in wet lab with many experimental assays. Each experiment could reveal a binary interaction or multiple interactions; therefore a complete investigation requires the planning of a set of assays, under the guidance of some optimization strategy. Experimental techniques may aim to discover binary or multiple interactions, or they may investigate qualitatively or quantitatively the interactions themselves.

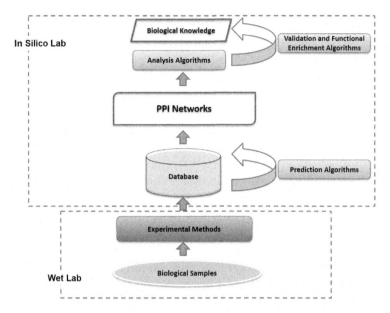

Figure 7.1. Flow of data and algorithms in interactomics. Starting from the bottom, biological examples are analyzed with experimental methods producing data related to interactions, stored in databases. Recursively these data may be used to add novel interactions to databases themselves. Then data are modeled using graph theory, and protein interaction networks are generated. Networks are then analyzed, and results may be validated through functional enrichment algorithms to support the discovered knowledge.

7.2.1 Experimental investigation of physical interactions

A set of techniques are aimed to determine the interaction of two or more proteins without looking at the dynamics of the reactions. Usually, these techniques are based on a *bait-prey* model. Therefore a *bait* protein is used as a test to demonstrate its relations with one or more proteins, called *preys*. These models have been employed in many experiments [21,35,41,98].

Protein chips

DNA microarrays have been used for the analysis of gene expression in many research projects. Microarrays are based on small chips organized into a rectangular array. Each spot of the array binds a different gene. There exist different microarrays that are used to analyze both quantitatively and quantitatively genes in a sample. The large use of these microarrays has suggested the possibility to realize the construction of chips for the building of spotted proteins instead of DNA, called protein chips or protein Microarrays [39,45].

Similar to DNA microarrays, this kind of arrays analyze thousands of molecules simultaneously. Historically, the first work on protein arrays [11] presented the screening of cDNA libraries searching for clones protein in *E.coli*.

The investigated arrays contained thousands of different expression clones bounded to protein-binding membranes. Moreover, it is remarkable to note that technology could be used to screen protein-protein interactions. In a research [105], 39 proteins interacting with calmodulin were identified, starting from the yeast proteome. All the 33 interacting proteins missed in the yeast-two-hybrid or the mass spectrometry was a novel prediction, and a new consensus binding site was defined. From a technological point of view, all arraybased techniques described below share the same principle: a set of investigated compounds are immobilized over a surface, and then the interacting partners are used as assays to demonstrate a specific interaction. The current state-of-the-art of protein microarray includes three kinds of arrays: (i) protein microarrays (PMA), which use antibodies or DNA to study functional characteristics of immobilized proteins, (ii) antibodies microarrays (AMA), which use purified proteins to characterize specific immobilized antibodies, and (iii) reverse protein microarrays [13] (RPMA), in which fractionated proteins, or complex mixtures are immobilized, and interactions with serum is investigated to profile the serum itself. For further details, see [18], [39].

The main drawback of this technique is in the preparation step that requires a high-quality expression library (the interacting partners) and an array production yielding a huge number of active proteins. Recently, various efforts aim to integrate the classical methodic into a high-throughput and automated system. Nevertheless, this field presents a lack of standardization methods both in the experimental setup and the data presentation and modeling.

Mass spectrometry

Mass spectrometry (MS) is a widely used analytical technique that measures the mass of ions by revealing the mass-to-charge ratio. Results of MS experiments are represented by a mass-spectrum, i.e., a two-dimensional function that measures the intensity as a function of the mass-to-charge ratio. Therefore a spectrum gives information about the presence and the relative abundance of an ion in a sample.

In a MS experiment, a sample is ionized, through the use of electrons. Then the molecules of the sample are breached into small ion fragments that are separated according to their mass-to-charge ratio. The separation may occur, for instance, by accelerating ions by using an electric or a magnetic field. Accelerated ions are detected by a mechanism that can detect these charged particles. MS can detect known molecules, since each molecule has a characteristic spectrum, and it can detect unknown molecules by querying some databases. The use of MS in interactomics [21] is based on the use of a bait protein and the selective separation from the protein, and its interactors from a cell lysate. The set of bait and preys are digested into peptides using many enzymes, and then the sample is analyzed through mass-spectrometer. The resulting mass spectra contain the signature of all the interactions. MS experiments identify protein complexes, not pairwise interactions, which are conversely identified by yeast-two hybrid assays.

The core of MS is the isolation of protein complexes from the other compounds through purification methods, which use commonly tagged target proteins. Many different experimental methods have been designed to analyze a high number of proteins. Usually, a c-DNA clone is used to produce a tagged bait protein, then cells are transfected, and prey proteins bind the bait. The complexes of the tagged proteins are finally purified and analyzed by MS [83]. A comprehensive dissertation on this method and a comparison concerning yeast-2-hybrid can be found in [21].

Yeast-two-hybrid

The yeast-two-hybrid (Y2H) [22,23] system is a largely used technique for the detection of protein interactions. Y2H is based on the simulation of the interaction in yeast and the subsequent prediction in other organizms; therefore the reliability is lower than other techniques, such as the previously described.

The assays are based on the use of the transcription factor GAL4. A transcription factor is a protein that regulates the activation of transcription in the eukaryotic nucleus. Galactose activates the transcription of GAL genes and the production of the galactose metabolic protein having two domains, an activation domain and a DNA-binding domain. The bait protein binds the GAL4 activation domain, whereas the prey binds the GAL4 DNA-binding domain. If the two proteins interact, the reconstituted factor will activate the transcription of a reporter gene, which has been engineered to contain the GAL4 promoter. Nevertheless, the determination of interaction with this assay is only a possibility that two proteins interact *in vivo*. To get over this difficulty, the interaction determined *in vitro* are accepted with more confidence if they share the biological process or the cell compartment.

Many works ([27,56,82]) demonstrate the use of this techniques to generate large datasets.

Surface plasmon resonance

Surface plasmon resonance (SPR) [5,40,93] is a technique used to study biomolecular interactions, e.g., antigen-antibody binding. The first step of this technique involves the immobilization of all the biomolecules capable of binding to specific analytes or ligands on one side of a metallic film. Then the light is focused on the opposite side of the film. In such a way, the light excites the surface plasmons, that is, the oscillations of free electrons propagating along the film's surface. Finally, the refractive index of light reflecting off this surface is measured. The changes in this index measure the binding of immobilized biomolecules and their ligands. Clearly, an alteration in surface plasmons on the opposite side of the film is created that is directly proportional to the change inbound, or absorbed, mass.

When the affinity of two ligands (i.e., bait and prey) has to be determined a so-called bait ligand is coated to the surface of the crystal. Through a microflow system, a solution with the prey ligand can flow over the bait layer and bind. Binding modifies the SPR signal until an equilibrium is reached. Thereafter, a solution without the prey is applied, and a new equilibrium is reached. The binding constant can be calculated by comparing the two changes in the signal.

7.2.2 *In silico* inference of PPI

Finding physical interactions is always not feasible, due to the expensive experimental setup. At the same time, it is time-consuming. Experimental methods are prone to incompleteness and noise. The best alternative is to guess computationally the possible physical interaction based on available data sources. Computational methods for predicting interactions among protein molecules *in silico* are broadly classified into two categories.

Similarity-based methods

Considering different properties of proteins, such as sequence homology, gene coexpression and phylogenetic profiles [53,70, 87], pairwise similarity is computed between a pair of proteins to predict a possible interaction between them. In addition to nonstructural information, structural data about a pair of proteins appears to be more effective in improving prediction accuracy [91,104]. Performance of any similarity-based approaches is sensitive to the quality of data used for prediction and the merit of the similarity measures used for the same. In reality, predicting deterministically whether two given proteins are physically interacting or not, based on the similarity of different structural and nonstructural features, is a challenging task due to the above crucial factors. Few researchers recently address the issue by improving the reliability of the input source with the help of integrating heterogeneous features, including genomic and semantic features [17,38,42]. Researches are also extended towards predicting the link between a pair of proteins with the help of network topology.

Network-based methods

An initial seed network can be constructed based on few of the reliable pairwise PPIs, with nodes representing proteins and edges representing interactions. Network topological features, such as the degree of a node, are used as a similarity for predicting the link between any pair of proteins (nodes) [55,80]. Alternatively, random walk-based methods have been explored, involving the entire network. Two nodes, possibly connected to each other if a random walk started in one node, is likely to reach the other node [97]. Spectral analysis [95] and matrix factorization [102] based solutions were also provided through various researches.

Both types of methods discussed are complementary to each other. Whereas similarity-based can be used to assign a weight of a link, random walk can be applied to the weighted network to find more likely walk between the nodes [7]. Heterogeneous multimodal sources are also used for inferring the possible links with the help of canonical correlation analysis [103].

7.3 Fitting protein networks into models

The determination of a correct model for PPI networks could be important for effective experimental planning, helping to determine possible interactions. Currently, there exist five common models used for PPI network, as discussed before: (i) random graph model, (ii) scale-free model, (iii) geometric random graph model, (iv) STICKY model and (v) hypergeometric graph model.

In network theory, the individuation of a theoretical model, to which network belongs may provide relevant information, i.e., as a tool for prediction of novel interactions, and for automated analysis. For instance, Lappe et al. [54], used a theoretical model to suggest correct experiments for the individuation of the human PIN to reduce the total number of experiments.

The development of a general model for interaction networks can capture relevant information and can provide a tool for predictive inference. The introduction of a common shared model is currently a research area. The simple representation of interaction datasets as a graph does not consider the following problems as noted in [77]: (i) it does not take into account a set of supplementary information about the interactions, such as spatial and temporal information, kinetics parameters, number of experiment confirming the interaction (see [101]), and (ii) the large number of false-positive and false negative interactions present in the dataset. For instance, when a physical interaction is detected as direct interaction, in true sense it happens only in the presence of mediators.

Conversely, a correct model of PPI network enables researcher to build predictive experiments (i.e., predicting the possible interaction, and hence verifying them with experiment), to analyze structural properties (e.g., individuation of complexes or pathway) and to bring together information about the function of themselves.

As described in [54], for the human PPI network, a model has been used to plan high-throughput experiments. In that work, a theoretical model is used to guide experiments, thus avoiding unuseful experiment in classical strategies. In a typical experiment, a so-called pull-down strategy is adopted. This strategy involves choosing a single protein that plays the role of a *bait* and identifying all interacting partners of this protein without any optimization strategy.

It is evident that an incorrect model can suggest experiments that do not reveal any interaction. In such situations, the map is bigger than the empire or can fail to indicate real interaction, which will remain hidden. Despite this, it seems reasonable that due to small coverage of PPI data with respect to the interactome, once a complete dataset is available, this could be used to evaluate existing models.

Molecular machinery inside cells often involves the interplay of molecules of different types (e.g., genes, proteins, and ribonucleic acids [29]). Consequently, the use of more complex models, such as heterogeneous networks, to represent such data has gained the attention of researchers. Heterogeneous networks use nodes and edge of different types. The use of heterogeneous networks has been explored for data integration. Przytycka et al. [81] explored integrating different types of molecules (genes, proteins, and transcription factors) and their various kinds of interactions into a heterogeneous network. Mitra et al. [69] discussed a lot of these approaches in a review, and the recent study by Cowen et al. [16] summarizes all these approaches.

7.3.1 Searching the best fitting model

One big challenge of research in biological networks is the individuation of the best theoretical model, to which real network belongs to. The individuation of this model is based on two main steps: (i) the individuation of some *robust* properties of the networks that may summarize the whole network (e.g., node degree distribution, clustering coefficient distribution); (ii) the comparison of these properties with respect to those of the model networks having same nodes and edges. Common statistical properties of the networks are the degree distribution, clustering coefficients, node centrality distribution, and network diameter. These properties require, in general, relatively low time for the calculation, but unfortunately, they are significantly affected by noise. Consequently, other local properties more robust to noise have been introduced, as described in what follows.

Graphlet-based models extract small subgraphs from the large networks. Graphlets are particular subgraphs of a graph $\mathcal{G}$ defined as follows.

Given $\mathcal{G} = (\mathcal{V}, \mathcal{E})$, $|\mathcal{V}| = n$, $|\mathcal{E}| = m$, a k-graphlet is a subgraph of $\mathcal{G}$ composed of k connected nodes [79], and usually k ranges in the interval 2–5. Each k graphlet has a set of automorphism orbits. Therefore, it is possible to enumerate all the graphlets on 2, 3, 4, and 5 nodes. Given a node v of a graph, it is possible to count the number of each graphlet orbit to which it belongs. The vector of these counts is also called the graphlet-orbit signature.

Przuly et al. in [77] investigate the structure of two available networks of *Saccharomyces cerevisiae* and *Drosophila melanogaster*. Starting from these networks, the authors identify all the possible graphlet of dimension 3–5. Then they compare the frequency of the appearance of these structures with respect to the frequency measured in four types of model networks: (a) Erdös-Rényi (ER) networks with the same number of nodes and edges as the original PINs; (b) ER networks with the same number of nodes, edges, and the same degree distribution as the original PINs; (c) scale-free random networks with the same number of nodes and the number of edges within the original PINs; and (d) geometric random graphs with the number of nodes and the number of edges within those of the original PINs. Using this new measure of network structure, authors conclude that the geometric random model is the best one.

More recently, Maharaj et al., [58], extended previous results about model fitting. They evaluated different models of networks (some are reported in Table 7.1) on nine interactomes of different species, stored into the BioGRID database. Similar to previous studies, they find that STICKY (stickiness index-based) model is usually the overall best-fitting model for all nine networks. The main contribution of this work is the finding that other models, e.g., the hyperbolic geometric and scale-free models sometimes are better for some networks and that the geometric random graph model, which was previously found to be good as the STICKY model, is now roughly tied with ER as the worst-performing model.

Table 7.1 Theoretical models for PINs.

Name	Description
Erdos–Renyi (ER)	The probability of the existence of an edge for each pair of nodes (u, v) is p.
Scale free (SF)	The degree distribution asymptotically follows power law. There exist few nodes with high degree, (hub nodes).
Geometric random graph (GEO)	Nodes are points in a metric space. There exist an edge between two nodes u, and v if they are within the radius of each other.
STICKY model	Nodes assigned stickiness index based on relative degree according to input degree distribution. Product of stickiness indices between nodes determine the probability of the existence of an edge.
Hypergeometric graph model (HGG)	Nodes are points on a hyperbolic disk. The probability of the existence of an edge between two nodes is a function of their distance in hyperbolic space.

7.4 PPI network alignment for comparison

The current abundance of PPI networks is helping to address the challenge of discovering conserved interactions across multiple species. Comparison of conserved substructures across species helps one to understand complex biochemical processes. The ultimate goal of network alignment is to predict the protein functions of an unknown species from known ones. Sequence similarity metrics, such as BLAST scores, do not provide conclusive evidence for similar protein functions. PPI network alignment can be treated as a supplement to the computation of sequence similarity, providing topological information that may be needed to identify orthologous proteins. Protein network alignment is a relatively young research area, and successes of PIN network alignment, so far, include uncovering large shared subnetworks between species as diverse as *S. cerevisiae* and *H. sapiens*, and reconstructing phylogenetic relationships among species based solely on the amount of overlap discovered between their PPI networks [51,52]. Comparing two or more biological networks is a challenging problem, since many interesting questions we might ask of these networks are computationally intractable to answer [6]. Assuming we have k graphs and each graph has n nodes (in reality, it is likely that the graphs have an unequal number of nodes), each matching tuple a_i contains k nodes, one from each graph. Ideally, we should have n such tuples in the overall alignment A with the stipulation that an alignment metric value is optimized. This is a very complex optimization problem, which is NP-complete, and thus, all approaches have to be heuristic. Most papers in the literature report promising results in creating alignments that show either large regions of biological, or of topological similarity between the PPI networks of various species, but few do both well.

Network alignment is based on graph comparison through graph or subgraph isomorphism and may be seen as the counterpart of sequence analysis for proteins [31,66]. Since the alignment problem relies on the *(sub)graph isomorphism problem*, which is computationally hard in some general formulations [49], all the existing algorithms employ ad hoc heuristics. Formally, the problem of graph alignment may be defined as follows:

Definition 7.4.1 (PPI Network alignment). Given k distinct PPI networks $\mathcal{G}_1 = \langle \mathcal{P}_1, \mathcal{I}_1 \rangle$, $\mathcal{G}_2 = \langle \mathcal{P}_2, \mathcal{I}_2 \rangle$, $\cdots$, $\mathcal{G}_k = \langle \mathcal{P}_k, \mathcal{I}_k \rangle$ from k different species, the PPI alignment problem is to find a conserved subnetwork, A, within each of the k graphs. The alignment graph $\mathcal{A}$ is a subgraph consisting of nodes representing k similar proteins (one per species), and edges representing conserved interactions among the species. Each alignment $\mathcal{A}_i$ can be represented as

$$\mathcal{A}_i = \{\langle p_{1i}, p_{2i}, \cdots, p_{ki}\rangle | p_{ji} \in \mathcal{P}_j\}. \tag{7.1}$$

The alignment problem looks for optimal set $\mathcal{A}(\mathcal{G}_1, \cdots, \mathcal{G}_k) = \{A\}_{i=1\cdots k}$ alignments by maximizing the coverage of vertices from all the participating networks.

Existing alignment methods may be grouped into two major categories on the basis of their strategy: local network alignment (LNA), and global network alignment (GNA). LNA algorithms search for relatively small similar subnetworks in the (two or more) input networks, which may represent conserved functional structures [34]. Such algorithms were initially developed for the individuation of common building blocks of cellular functions [51] representing conserved machinery during the evolution. The GNA algorithms try to determine the best superimposition of the whole input networks matching all the input nodes. GNA search a comprehensive mapping of all the nodes, and they usually are used to transfer knowledge among networks [51]. An illustrative view of the two types of alignments are shown in Fig. 7.2A and Fig. 7.2B. Recently, some different approaches are proposed, based on a possible integration of LNA and GNA for improving the quality of results [59].

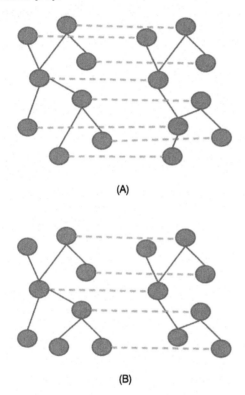

(A)

(B)

Figure 7.2. Network alignment approaches. (A) Global network alignment maps all possible proteins of one network to other. (B) Local alignment align subset of nodes with best matching scores.

Both LNA and GNA may also be categorized on the basis of the input networks in **pairwise alignment algorithms**, i.e., algorithms that align two input networks, and **multiple alignment algorithms**, i.e., algorithms aligning three or more input networks at once.

7.4.1 Pairwise alignments

A pairwise PPI network alignment is an alignment limited to only two species, i.e., $k = 2$. Given two networks, it finds an optimal matching of nodes between the two. The main objective of aligning two PPI networks from two species is to find conserved proteins in the two species [75,96]. Most published work on network alignment focuses on pairwise alignment. Given two PPI networks, one can perform two types of alignments: global alignment or local alignment.

In global alignment, each node of one network is aligned to one and only one node in the other, producing a single overall match of two or more networks. In global network alignment, each node in the smaller network is uniquely aligned to only one best matching node in the larger network. Such an alignment detects a maximal subnetwork, which is useful in functional orthology prediction.

In contrast, local alignment attempts to find the best alignment of subnetworks of one network with subnetworks of another, without worrying about how other nodes line up. Local alignment methods may produce one-to-many node alignments, where a node from one network may be aligned to more than one node from the other network. In local alignment, small regions of similarity are matched independently, and several such regions may overlap in conflicting ways. Computationally, local alignment is less expensive than global, as global alignment aims to find a single large consistent match.

Local network alignment algorithms

As discussed above the *LNA* algorithms aim to find multiple and unrelated regions of isomorphism between the input networks, where each region implies a mapping independently of the other ones. The strategy results in a mapping or set of mappings between subsets of nodes such that their similarity is maximal over all possible subsets. These subnetworks correspond to conserved patterns of interactions that can represent a conserved motif or pattern of activities. The literature contains many ways to categorize LNA algorithms [68]. LNAs can be broadly categorized into two main classes on the basis of the overall strategy as depicted in Fig. 7.3. Algorithms belonging to the first class, named *mine and merge*, first analyze each network separately, and then project solutions reciprocally from a PIN to the others. In the merge step, the input networks are integrated into a single alignment graph on the basis of their topology and on supplementary information provided. Then this graph is analyzed, and local alignments are extracted. Algorithms belonging to the second class, *merge and mine*, i.e., algorithms that first integrate PINs into a single graph, and then analyze such graph. Every single network is analyzed, usually by using a community discovery al-

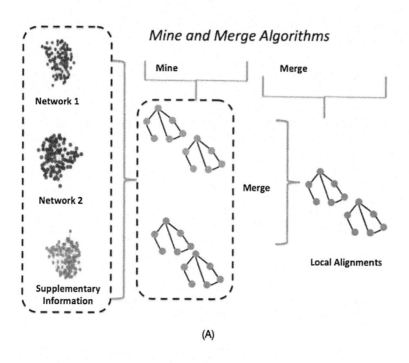

(A)

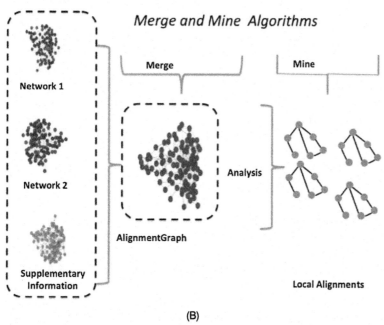

(B)

Figure 7.3. Two common approaches adopted in local alignments. (A) Merge and mine approach. (B) Mine and merge approach.

gorithms. Then all the pairs of communities of the input networks are matched, and alignments are generated.

Mine and merge analysis are usually less expensive in terms of computational resources, as evidenced by Erten et al. [20]. In general, merge and mine algorithms are more complicated, due to difficulties in formulating and accounting for approximate matches, and the existence of multiple mappings between proteins in different species [20]. Moreover, they are computationally expensive, since, in order to merge the input networks, it is necessary to compare their topologies. The main drawback of these approaches is that these algorithms are more sensitive to the noise in input networks and to the redundancy of information in input networks. Conversely, merge and mine algorithms are less sensitive to these problems, but they present, in general, a higher computational cost in the building of the initial integrated graph (also referred to as alignment graph). The interested reader may find a detailed discussion of these approaches in Mina and Guzzi [68].

A common problem in both groups is the requirement of additional information used to seed to build the alignment. Such seeds are ortholog pairs, and the absence of such information would require the exploration of all the possible combinations of protein pairs. Consequently, many algorithms require as input two networks and a list of protein pairs (for instance, a list of putative orthologs) to start the computation. Lists of pairs may be obtained from existing databases of orthologs, or gathering sequence similarity information using tools, using semantic similarity [30]. In what follows, we discuss the next few popularly used network aligners. A comprehensive discussion of different aligners can be found in [84].

PathBLAST   PathBLAST [46] is a local protein-protein aligner and is one of the earliest PPI alignment methods. Just as BLAST is used to align protein sequences, PathBLAST aligns protein networks. Interacting pathways, which are chains of proteins, are aligned with a pathway from the second network, in which the proteins occur in the same order. In this way PathBLAST restricts itself to linear chains of interacting proteins, simplifying the problem of network protein alignment. The alignments in PathBLAST are scored based on the similarity of the protein sequences at each pathway position, the quality of the protein interactions. PathBLAST allows gaps in the other network pathway alignments, i.e., where interacting proteins in one path can be aligned with a path in the other network, where the proteins are not directly connected but instead are connected by an intermediate protein, i.e., at a distance of two. A web-based tool for PathBLAST is available[1] for seven species, where protein ID or protein sequences are provided by the user. PathBLAST reports high-scoring alignments of the query network and the network on one of the available species.

NetworkBLAST   NetworkBLAST [44] is a local protein-protein interaction network aligner. Unlike PathBLAST, which detects paths, NetworkBLAST detects complexes that are conserved through evolution, between two species. It can also find complexes in a single network or species. A web-based tool[2] is available for NetworkBLAST, which takes two PPI networks as input, and the BLAST E-value of pairs of proteins. In the case of a single species, only the PPI network is required as input. NetworkBLAST predicts complexes based on the density of the subnetworks, using a log-likelihood ratio score, and nodes with high scores in the subnetworks are identified and used as seeds in a greedy expansion procedure.

NetAligner   NetAligner [73] is another local network aligner, which finds conserved interactions of proteins based on evolutionary distances. From two input PPI networks, NetAligner creates a complex protein network based on the proteins that are present in the networks and the interactions from the interactomes of the species. Seeds are identified from the connected components detected from the initial alignment and then are extended to other seed vertices using gaps or mismatch edges. In this extended alignment graph, the connected components are the final alignment. NetAligner is available online,[3] where the user can select from three categories: complex to interactome, pathway to interactome, or interactome to interactome. In the complex to interactome mode, the query is a complex, which is mapped to an interactome of the same, or a different species. In the pathway to interactome mode, the query is a pathway from a species, which is mapped to the interactome of a different, or the same species. This helps detect pathways that are conserved across species. In the interactome to interactome mode, whole networks or interactomes are mapped to each other to find conserved complexes between the species.

MaWISh   MaWISh [50] is an early PPI aligner. It is a local PPI aligner that predicts conserved complexes, which are defined as

[1]http://pathblast.org/.
[2]http://www.cs.tau.ac.il/~bnet/networkblast.htm.
[3]http://netaligner.irbbarcelona.org/.

locally maximal heavy subgraphs. In an alignment graph, conserved complexes that have similar functions are highly connected and dense, with fewer interactions with other complexes. MaWISh uses a greedy approach, which finds conserved hub nodes and extends them to nodes, which have edges in both networks. MaWISh uses the min-cut algorithm to find a minimum number of edges in which to partition the graph, maximizing the weight of internal nodes to extract subgraphs. In MaWISh, balance is not considered, and the weight of only a part of the graph is considered. This method iteratively produces new heavy subgraphs until the subgraph becomes redundant (when r% of nodes are present in subgraph). In this way, locally maximal heavy subgraphs, corresponding to or conserved complexes, are extracted from the PPI networks.

AlignNemo   AlignNemo [14] considers both biological function and topology of interactions during alignment. It is able to uncover subnetworks of proteins that relate in biological function and topology of interactions. The algorithms are able to discover conserved subnetwork with a general topology without referring to specific interaction patterns. Conserved modules are, for instance, functional complexes. The algorithm is able to handle sparse interaction data with an expansion process that at each step explores the local topology of the networks beyond the proteins directly interacting with the current solution.

Align-MCL   AlignMCL [68] is an evolution of AlignNemo. In particular, AlignMCL builds the local alignment by merging all the input data in a single graph, named *alignment graph*, which is afterwards examined, and by using the MCL, Markov cluster algorithm, to extract the conserved subnetworks [19], without the imposition of any particular topology. During the building of the alignment graph (see [14] for complete details), AlignMCL scores the link between two nodes of the alignment graph by estimating the number of paths of length ≥ 2 connecting the two nodes in the original networks. A strategy based on the Jaccard index is applied to weight each score opportunely. In this way, the scores take into account the node degree of input networks. Then, the final alignment graph undergoes a pruning step that locally removes the weak links, considerably reducing the network. In the next step, AlignMCL applies the MCL clustering algorithm to the alignment graph.

LocalAli   LocalAli [37] uses maximum-parsimony evolutionary model for the building of local alignment among multiple networks as functionally conserved modules. LocalAli uses the maximum-parsimony evolutionary model to infer the evolutionary tree of networks nodes. Then, LocalAli extracts local alignments as conserved modules that have been evolved from a single ancestral module.

Global network alignment algorithms

The *global network alignments, GNAs,* aim to find a mapping that should cover all of the nodes of the input networks, associating each node of a network with one node of the other network or marking the node as a gap when no possible match exists. This strategy tries to find a consistent mapping between the whole set of nodes of the networks.

There exist many GNA algorithms, and here, we briefly present a little selection of them discussing only the most recent ones.

IsoRank aligners   The IsoRank algorithms [92], [57], are the earliest developed global aligners. The basic IsoRank algorithm has two major versions. Whereas the first version deals with global pairwise network alignment, the second version of IsoRank solves the problem of global multiple network alignment. The main idea of the IsoRank algorithm is to use an algorithm similar to *Google's PageRank* [74] to estimate node similarity, by recursively defining node similarity in terms of the similarity of a nodes' neighbors.

IsoRankN   IsoRankN [57] or *IsoRank-Nibble* is an enhancement to the original IsoRank, presented earlier. It solves the problem of global multiple network alignment and can be used as a global pairwise aligner when only two networks are given as input. IsoRankN also works in two stages as in the original IsoRank algorithm. In the initial step, a similarity matrix is calculated for each pair of input networks. This step can be accomplished by running the original IsoRank algorithm on each pair of networks without performing the second round of IsoRank, where the final alignment is produced. The final alignment is generated in the second phase of IsoRankN by using an iterative spectral clustering algorithm [4]. The standard output format for IsoRankN and all global multiple alignment programs is two sets of nodes, one from each network, which are aligned to each other. This format makes it difficult to evaluate the alignments according to the evaluation metrics [71]. We note that the original implementation of IsoRank can align, at most, 5 networks of various species, due to its exponential time complexity; the time for aligning more than six networks with IsoRankN is extremely high and therefore impractical [36].

Moreover, the iterative nature of IsoRankN, and the need to update large matrices at each iteration, makes IsoRankN worthless in aligning large networks, particularly those that have more than 10,000 nodes [90].

MAGNA MAGNA (maximizing accuracy in global network alignment) [88] is a GNA that uses a genetic algorithm to build a high-quality alignment, improving a set of existing ones (generated randomly or chosen using other aligners). In detail, MAGNA simulates a population of alignments that evolves by applying the genetic algorithm and a function for the crossover of two alignments into a superior alignment.

MAGNA++ is an evolution of MAGNA [100], which introduces important improvements. Whereas MAGNA maximizes edge conservation during the alignment, MAGNA++ enables both the maximization of any different measures of edge conservation and any desired node conservation measure.

SANA Similar to MAGNA++, the SANA (simulated annealing network aligner) [62], is a global aligner based on simulated annealing. It takes as input two networks and an initial alignment; then, it starts to explore a solution space to improve the initial alignment. The solution space consists of alignments neighbors, i.e., alignments that vary only in one or two mappings of individual pairs of aligned nodes with respect to the initial alignment. In the end, the alignment that maximizes the objective function represents the final alignment.

GRAAL The GRAAL series is a collection of network aligners that use topological similarity, as estimated by graphlet degree signatures [79], to align one network to another. The intuition behind this methodology is that the local topology around the proteins in their respective protein networks is exhibited in the different ways proteins interact. Hence, homology information is deemed to be encoded in the network topology in addition to sequence similarity [51]. The GRAAL family consists of four different algorithms, namely GRAAL [51], H-GRAAL [67], MI-GRAAL [52], C-GRAAL [63], and L-GRAAL [60].

Global versus local alignments

LNA and GNA are clearly connected, since both aim to find topological and functional similarities among networks to transfer of biological knowledge from well-studied species to poorly studied ones. In the recent past, researchers have produced many independent GNA and LNA algorithms that rely on different assumptions and algorithmic solutions.

Consequently, a simple comparison of an LNA method and a GNA method is nontrivial, and when a new method is proposed, it is only compared against the same category. Therefore the ultimate question is which one to use: LNA, GNA, or a hybrid approach that would reconcile the two?

Guzzi and Milenkovic provided a comprehensive evaluation study [32], which showed that LNA and GNA are complementary. The former results in high functional but low topological quality, whereas the latter results in high topological but low functional quality, and since they lead to very different biological predictions. Authors conclude that a possible reconciliation between these two aspects of network alignment is an open research problem that should be investigated deeply in the future.

The IGLOO (integrating global and local biological network alignment) [64] integrates algorithmic aspects of GNA and LNA to build alignment with high functional and topological quality. Initially, IGLOO uses an LNA approach to build local alignment, as a set of small conserved subnetworks of high functional quality. Then, IGLOO computes a new node cost function (NCF) resulting from the combination of node topological similarity (TS), node sequence similarity (SS), and interaction score (IS). In a dual fashion, the GLAlign algorithm [65] improves the performance of local network aligners by exploiting a preliminary global alignment.

7.4.2 Multiple alignment

Optimally one PPI network to another is NP-complete, and that is why all algorithms for 2 network mapping are heuristic. If we have k PPI networks (say $k = 50$ through a few thousand) and are tasked with finding an alignment of all the networks, an already difficult problem becomes more difficult. Aligning multiple PPI networks may enable a better understanding of protein functions and protein interactions, leading to a better understanding of the evolution of species. Though multiple network aligners have been designed and implemented to deal with more than two input networks, they can still be used as pairwise aligners when the number of PPI networks is only two. Below, we provide a brief sketch of some of the currently available multiple aligners.

Græmlin Græmlin [24] is one of several algorithms that align multiple networks and is unique in requiring phylogenetic information as input, in addition to network data. There are two versions of this algorithm. Whereas the first algorithm, known as *Græmlin*

1.0 [25], performs local alignment, the second version of Græmlin is a global aligner, known as *Græmlin* 2.0 [24]. Græmlin is one of the earliest aligners and works in two phases. In the first phase, a feature-based pairwise scoring function is calculated as a vector of pairwise node and edge features. In the second phase, Græmlin updates the initial alignment results from the first step and continues to iteratively update the alignment until the final alignment is found.

SMETANA The goal of SMETANA [86] is to find the maximum expected alignment for large networks effectively. As with most aligners, SMETANA also uses two phases. In the first phase, the similarity score matrix, which shows the likelihood of aligning two nodes, based in the semi-Markov random walk model estimated using the *semi-Markov random walk model* [85]. The second phase of SMETANA finds the final alignment using the scoring matrices constructed and enhanced in the first phase.

NetCoffee NetCoffee [36] is another recently published global multiple aligner. It is an aligner that can be easily adjusted to align any kind of networks, not only PPI networks. NetCoffee is also one of the few network aligners available as an online tool. NetCoffee is unique among aligners in that it is a four-phase aligner. Another important feature that makes NetCoffee unique is that it combines biological and topological similarities to match two or more PPI networks. NetCoffee is the extension of a prior method, called *T-Coffee* [72], which is used for multiple sequence alignment. For each pair of networks to be aligned, a set of all bipartite graphs is built, known as a *bipartite graph library*. Then NetCoffee assigns weights to all edges using an integrated approach similar to *T-Coffee*. Sequence similarity is calculated as a log-ratio and integrated with topological similarity using an α parameter method.

BEAMS BEAMS [1] *(backbone extraction and merge strategy)* is a global multiple network aligner that combines both topological and biological similarity to calculate a multiple global alignments. The biological similarity between proteins is calculated based on BLAST bit scores, and balanced with topological similarity using the α parameter. BEAMS starts by constructing a k-partite similarity graph, where k is the number of input networks. The constructed k-partite weighted graph works as a superset, from which the final alignment is extracted. The final alignment is a subset, a *filtered version*, of the complete k-partite graph of the input networks, and contains only edges that have maximum weights. The step of constructing the similarity graph can be considered an off-line step, as it needs to be done only once in advance, and it does not affect the overall running time.

7.5 Protein networks complex detection

Instead of any single protein or gene, a group of proteins playing a synergetic role towards performing any functional activity. A function carried out in an organism ideally is an outcome of orchestral coordination that occurs among the set of proteins. Such a group of active proteins in an interaction network is termed as *protein complex*. Similar to that, a group of proteins may be associated with a common biological function [76], called a functional module, which do not necessarily interact at the same time. However, the difference between the two groups is not very clear, as the protein interactions hardly give any structural or physical information.

Given a network of proteins, well-known clustering techniques can be applied to detect such network complexes. Clustering refers to the process of grouping data objects into groups (both overlapping and not) of *similar* objects such that similarity of the objects within the same cluster is greater than the similarity of the object outside of the cluster. In particular, clustering, a PIN, refers to the grouping of proteins that share a large number of the interactions (or a similar interaction pattern) compared to the average of the network. Consequently, the results of clustering are two-fold: giving insight into the internal structure of the network and suggesting possible shared function among proteins, hopefully unknown. Clusters of proteins represent protein complexes or functional modules.

Finding complexs by simulated random walk

The Markov cluster algorithm [19,99] (MCL) finds clusters on a graph by simulating a stochastic flow, and then analyzing its distribution. A network can be represented as a collection of paths sharing a starting point that guides a certain number of random walks. In observing the walks, one can see a particular behavior on the resulting flow: when a walker reaches regions highly connected, then the latter will have a little probability of getting out. In this way, considering the evolution of the flow, walks will occur in the regions with many edges, and walks interregions will be more and more infrequent.

So, the MCL algorithm simulates a collection of random walks within the network and iteratively it weakens the flow, where it is weak and increases the flow where it is strong (in the highly connected regions). This process will cause aparition of a cluster structure, and it will end when a set of regions with the flow are separated by regions without flow.

This idea is implemented by building a stochastic matrix from the graph, and then by simulating a flow with some algebraic operations. Formally, let us consider a graph G and its adjacency matrix M_g; an associated Markov matrix is defined by normalizing all columns of M_g. Each value of this matrix represents the tendency of a node to be attracted by the other ones. Clearly, at the first step, each node is equally attracted by its neighbors. The evolution of the system, i.e., of the flow, is computed by calculating the next powers of this matrix. A theorem states that for any Markov matrix, the successive computation of powers causes the achievement of a state, in which each node is not attracted by a particular node. Regardless of this fact, regions highly connected try to hold the flow within them.

The algorithm enhances this behavior with an operation called *inflation*, which changes the matrix values in order to increase the probability to reach a node in highly connected regions. The inflation operation, based on an inflation parameter greater than 1, influences the cluster structure, the greater the inflation parameter, the greater the number of clusters.

Diversely from the classical algorithms, the MCL does not suppose a defined cluster structure, i.e., a fixed number of clusters. Currently, MCL is implemented for Linux platforms, and it is freely available on the author's website.[4]

Discovering complexes by finding dense regions

The molecular complex detection (MCODE) algorithm, described in the early work of Bader et al., [8], tries to find complexes by building clusters. The rationale of MCODE that takes in as input the interaction network, is the separation of dense regions based on an ad hoc defined local density. MCODE has three main steps: (i) node weighting, (ii) complexes prediction, (iii) post processing.

In its first stage, MCODE weights all vertices based on their local network density. The local area, in which density is calculated, is delimited by an ad hoc defined subgraph structure called k-*core*. A k-core of a graph is the central most densely connected subgraph with minimal degree k. Thus the core-clustering coefficient of a vertex v is the density of the highest k-core of the immediate neighborhood of v. Finally, the weight of a vertex is the product of the vertex core-clustering coefficient and the highest k-core level, *kmax*, of the immediate neighborhood of the vertex.

[4]http://micans.org/mcl.

The resulting weighted graph is given as input of the second stage. The algorithm, hence, starting from the highest weighted vertex, tries to span a region visiting vertices, whose weight is above a certain threshold, called *vertex weight percentage* (WWP). This step stops when no more vertices can be added to the complex, and it is repeated considering the next highest weighted network not already considered. Finally, the third step has to filter the complexes, which do not contain at least a k-core with $k = 2$.

The third stage is the post-processing. Complexes are filtered if they do not contain at least a 2-core (graph of minimum degree equal to 2). The algorithm has two main options that determine the characteristics of this phase: *fluff* and *haircut*.

The algorithm has two modes of execution: a *direct mode* (in which the search starts from a given node), and an *undirect mode* (in which the seed is randomly selected). MCODE is freely available on the author's website,[5] and there also exists a version that runs as a plugin for the Cytoscape software.

Complex prediction via clustering

The paper of Ul-Amin et al. [2] presents another approach based on clustering an interaction network to find complexes. The rationale of the algorithm is the building of a cluster as a dense region embedded into a sparse region. The algorithm is logically organized in five major steps: (i) initialization, (ii) termination check, (iii) selection of starting node, (iv) cluster growth and (v) output.

In the first step, the algorithm takes as input an undirected graph and initializes its main variables: *cluster density, cluster property, cluster ID*. The algorithm calculates the minimum value of density for each generated cluster, i.e., the ratio of the number of edges present in the cluster and the maximum possible number of edges in the cluster. The cluster property $cp_{n,k}$ of any node n, with respect to any cluster k of density d_k and size $\|N_k\|$, is the ratio between the total number of edges between the node n and each of the nodes of cluster and the product between the density and the size of the cluster d_k. The cluster ID, k is initialized to 1.

In the second step, the algorithm verifies the termination conditions, and if the graph has no edges, the algorithm will end. Conversely, if the termination check fails, the algorithm in its third step, namely *selection of starting node*, will select a node as a starting point to build a new cluster. Hence, in the fourth step, namely *cluster growth*, the algorithm adds nodes to the cluster chosen from the neighbors of starting node. Neighbors are labeled with

---

a priority in order to guide the cluster formation. Finally, when a cluster is generated, it is removed from the graph, and the cluster-ID, k is incremented.

The algorithm is polynomial, and its complexity in the worst case is $O(N^3)$, where N is the number of nodes. This complexity is due to the cost of sorting clusters.

Cost-based clustering

The algorithm described in the work of Przuly et al. [47] tries to find complexes by clustering protein interaction networks and filtering the generated clusters. In this scenario, clustering the network corresponds to its decomposition into different subsets of nodes, inducing dense subgraphs [78]. The algorithm proposed by the authors, after an initial random clustering, is the restricted neighborhood search (RNSC), a cost-based local search algorithm based on the tabu heuristic as defined in [28] (see [48] for a detailed description of the algorithm).

The RNSC assigns to each partition of nodes a cost and, hence, searches this space. The algorithm uses two cost functions: an integer-valued cost function (known as a naive function) and a real-valued function (called scaled function). The first function acts as a fast preprocessor and the second one assigns a more sophisticated cost on the initial network clustering, which can be random or user-input. Finally, the clusters generated are filtered. Small clusters are discarded for two reasons: (i) small known complexes generally have a low density in known PPI networks, (ii) the overlapping of a small predicted complex and a true complex has more probability to occur by chance.

Predicted clusters are, hence, matched to known biological complex if the predicted cluster is entirely contained in the complex, or it overlaps a large proportion of the complex.

Complex identification through chordal graphs

The methods before presented are based on the interpretation of a complex as a particular region in the graph, and this region has been defined on the basis of an ad hoc density measure. These regions can be interpreted as functional modules of the graph, that is, a set of components interacting for targeting a specific goal. The translation of such a concept in a graph property is the rationale of different approaches. The work in [106] defines a functional module as a union of overlapping functional groups, which are dense subnetworks. These groups are represented by a particular clique (both maximal or ad hoc defined) on the graph. The proposed method is linked to a possible decomposition of chordal graph called clique tree representation. Although not every protein interaction network is represented by a chordal graph, authors developed a framework which generalizes this representation and represents each complex as a node of a so-called *tree of complexes*. Each node of the tree is a functional group and, for every protein, all functional groups, which is included, form a single subtree. Moreover, depending on the nature of the input network, some temporal relations between functional groups can be revealed.

Biological application: individuation of protein complexes

As introduced by Hartwell [33], a functional module is a group of cellular components, to which can be attributed a specific biological function. Consequently, molecular interactions networks can be organized in a set of modules of a small number of participants, which are low interacting with other modules.

A protein complex is a group of two or more associated proteins, which interact, sharing the same biological goal. For example, the breast cancer protein 1 (BRCA1) is known to participate in multiple cellular processes by multiple protein complexes, such as in association with the BARD1 protein or with the proteins Rad50-Mre11-Nbs1 [15].

Starting from an interaction network, complexes may be identified by searching for small and highly interconnected regions, known as *cliques* [8]. Prediction of molecular complexes can provide a method for functional annotation above other guilt-by-association methods. Predicted complexes can be already known, i.e., their composition is known, or belonging to a new protein complex. In this case, if the experiments confirm this relation, the algorithms can be used as a predictor. Finally, completely new complexes can be predicted.

7.6 Summary

Modeling and analyzing association among molecules in living cells is one of the challenges in computational biology. Therefore the so-called analysis of *interactome*, i.e., the whole network composed of proteins and their interactions are one of the most prominent fields of research. Protein interactions may be considered as a baseline for validating and elucidating other genomic interaction. This is because protein interactions can easily be verified experimentally for confirming true physical interactions. We discussed graph or network alignment problem. While aligning two or more networks, it is necessary to align networks that satisfy both biological and topological similarity simultaneously, which most of the aligners failed to address. Multiple alignments is a computationally intensive task, and hence need more simultaneous alignment procedure in the future, involving a large set of nodes and also a larger number of input networks for simultaneous alignment.

Acknowledgment

Authors thank Ms. Hazel N Manners, Dr. Ahed Elmsallati, and Prof. Jugal Kalita for their collaboration while writing this chapter.

References

1. Ferhat Alkan, Cesim Erten, BEAMS: backbone extraction and merge strategy for the global many-to-many alignment of multiple PPI networks, Bioinformatics (2013) btt713.
2. Md Altaf-Ul-Amin, Yoko Shinbo, Kenji Mihara, Ken Kurokawa, Shigehiko Kanaya, Development and implementation of an algorithm for detection of protein complexes in large interaction networks, BMC Bioinformatics 7 (1) (2006) 207.
3. Stephen F. Altschul, Warren Gish, Webb Miller, Eugene W. Myers, David J. Lipman, Basic local alignment search tool, Journal of Molecular Biology 215 (3) (1990) 403–410.
4. Reid Andersen, Fan Chung, Kevin Lang, Local graph partitioning using pagerank vectors, in: 47th Annual IEEE Symposium on Foundations of Computer Science, 2006, FOCS'06, IEEE, 2006, pp. 475–486.
5. K. Aslan, J.R. Lakowicz, C. Geddes, Plasmon light scattering in biology and medicine: new sensing approaches, visions and perspectives, Current Opinion in Chemical Biology 5 (9) (2005) 538–544.
6. Nir Atias, Roded Sharan, Comparative analysis of protein networks: hard problems, practical solutions, Communications of the ACM 55 (5) (2012) 88–97.
7. Lars Backstrom, Jure Leskovec, Supervised random walks: predicting and recommending links in social networks, in: Proceedings of the Fourth ACM International Conference on Web Search and Data Mining, ACM, 2011, pp. 635–644.
8. Gary Bader, Christopher Hogue, An automated method for finding molecular complexes in large protein interaction networks, BMC Bioinformatics 4 (1) (2003) 2.
9. Albert-Laszlo Barabasi, Zoltan N. Oltvai, Network biology: understanding the cell's functional organization, Nature Reviews. Genetics 5 (2) (February 2004) 101–113.
10. Pascal Braun, Anne-Claude Gingras, History of protein–protein interactions: from egg-white to complex networks, Proteomics 12 (10) (2012) 1478–1498.
11. K. Bussow, D. Cahill, W. Nietfeld, D. Bancroft, E. Scherzinger, H. Lehrach, G. Walter, A method for global protein expression and antibody screening on high-density filters of an arrayed cdna library, Nucleic Acids Research 26 (1998) 5007–5008.

12. Mario Cannataro, Pietro H. Guzzi, Pierangelo Veltri, Protein-to-protein interactions: technologies, databases, and algorithms, ACM Computing Surveys 43 (1) (2010) 1.

13. S.M. Chan, J. Ermann, L. Su, C.G. Fathman, P.J. Uetz, Protein microarrays for multiplex analysis of signal transduction pathways, Naturalna Medycyna 10 (2004) 1390–1396.

14. Giovanni Ciriello, Marco Mina, Pietro H. Guzzi, Mario Cannataro, Concettina Guerra, Alignnemo: a local network alignment method to integrate homology and topology, PLoS ONE 7 (6) (2012) e38107.

15. D. Cortez, Y. Wang, J. Qin, S.J. Elledge, Requirement of atm-dependent phosphorylation of brca1 in the dna damage response to double-strand breaks, Science 286 (1999) 1162–1166.

16. Lenore Cowen, Trey Ideker, Benjamin J. Raphael, Roded Sharan, Network propagation: a universal amplifier of genetic associations, Nature Reviews. Genetics 18 (9) (2017) 551.

17. Yue Deng, Lin Gao, Bingbo Wang, ppipre: predicting protein-protein interactions by combining heterogeneous features, BMC Systems Biology 7 (2) (2013) S8.

18. Arnaud Droit, Guy Poirier, Johanna Hunter, Experimental and bioinformatic approaches for interrogating protein–protein interactions to determine protein function, Journal of Molecular Endocrinology 34 (2005) 263–280.

19. Anton J. Enright, Stijn Van Dongen, Christos A. Ouzounis, An efficient algorithm for large-scale detection of protein families, Nucleic Acids Research 30 (7) (2002) 1575–1584.

20. Sinan Erten, Xin Li, Gurkan Bebek, Jing Li, Mehmet Koyutürk, Phylogenetic analysis of modularity in protein interaction networks, BMC Bioinformatics 10 (1) (2009) 333.

21. A.C. Gavin, et al., Functional organization of the yeast proteome by systematic analysis of protein complexes, Nature 415 (2002) 141–147.

22. S. Fields, O. Song, A novel genetic system to detect protein-protein interactions, Nature 340 (6230) (July 1989) 245–246.

23. S. Fields, R. Sternglanz, The two-hybrid system: an assay for protein-protein interactions, Trends in Genetics 10 (1994) 286–292.

24. Jason Flannick, Antal Novak, Chuong B. Do, Balaji S. Srinivasan, Serafim Batzoglou, Automatic parameter learning for multiple network alignment, in: Research in Computational Molecular Biology, Springer, 2008, pp. 214–231.

25. Jason Flannick, Antal Novak, Balaji S. Srinivasan, Harley H. McAdams, Serafim Batzoglou, Graemlin: general and robust alignment of multiple large interaction networks, Genome Research 16 (9) (2006) 1169–1181.

26. Joseph Gillen, Aleksandra Nita-Lazar, Experimental analysis of viral–host interactions, Frontiers in Physiology 10 (2019).

27. L. Giot, J.S. Bader, C. Brouwer, et al., A protein interaction map of drosophila melanogaster, Science 302 (5651) (2003) 1727–1736.

28. F. Glover, Tabu search, part I. Orsa, Journal on Computing 1 (1989) 190–206.

29. Pietro H. Guzzi, Maria Teresa Di Martino, Pierosandro Tagliaferri, Pierfrancesco Tassone, Mario Cannataro, Analysis of mirna, mrna, and tf interactions through network-based methods, EURASIP Journal on Bioinformatics and Systems Biology 2015 (1) (2015) 4.

30. Pietro H. Guzzi, Marco Mina, Concettina Guerra, Mario Cannataro, Semantic similarity analysis of protein data: assessment with biological features and issues, Briefings in Bioinformatics 13 (5) (2012) 569–585.

31. Pietro Hiram Guzzi, Tijana Milenković, Survey of local and global biological network alignment: the need to reconcile the two sides of the same coin, Briefings in Bioinformatics 19 (3) (2017) 472–481.

32. Pietro Hiram Guzzi, Tijana Milenković, Survey of local and global biological network alignment: the need to reconcile the two sides of the same coin, Briefings in Bioinformatics 19 (3) (2018) 472–481.

33. L.H. Hartwell, J.J. Hopfield, S. Leibler, A.W. Murray, From molecular to modular cell biology, Nature 402 (10) (1999) C47–C52.

34. Eitan Hirsh, Roded Sharan, Identification of conserved protein complexes based on a model of protein network evolution, Bioinformatics 23 (2) (2007) e170–e176.

35. Yuen Ho, Albrecht Gruhler, Adrian Heilbut, Gary D. Bader, Lynda Moore, Sally-Lin Adams, Anna Millar, Paul Taylor, Keiryn Bennett, Kelly Boutilier, et al., Systematic identification of protein complexes in saccharomyces cerevisiae by mass spectrometry, Nature 415 (6868) (2002) 180–183.

36. Jialu Hu, Birte Kehr, Knut Reinert, Netcoffee: a fast and accurate global alignment approach to identify functionally conserved proteins in multiple networks, Bioinformatics (2013) btt715.

37. Jialu Hu, Knut Reinert, Localali: an evolutionary-based local alignment approach to identify functionally conserved modules in multiple networks, Bioinformatics 31 (3) (2014) 363–372.

38. Lei Huang, Li Liao, Cathy H. Wu, Inference of protein-protein interaction networks from multiple heterogeneous data, EURASIP Journal on Bioinformatics and Systems Biology 2016 (1) (2016) 8.

39. C. Hultschig, J. Kreutzberger, H. Seitz, Z. Konthur, K. Bussow, H. Lehrach, Recent advances of protein microarrays, Current Opinion in Chemical Biology 10 (2006) 4–10.

40. E. Hutter, J. Fendler, Exploitation of localized surface plasmon resonance, Advanced Materials 16 (19) (2004) 1685–1706.

41. T. Ito, T. Chiba, R. Ozawa, M. Yoshida, M. Hattor, Y. Sakaki, A comprehensive two-hybrid analysis to explore the yeast protein interactome, Proceedings of the National Academy of Sciences of the United States of America 98 (2001) 4569–4574.

42. Woo-Hyuk Jang, Suk-Hoon Jung, Dong-Soo Han, A computational model for predicting protein interactions based on multidomain collaboration, IEEE/ACM Transactions on Computational Biology and Bioinformatics 9 (4) (2012) 1081–1090.

43. Hawoong Jeong, Sean P. Mason, A-L. Barabási, Zoltan N. Oltvai, Lethality and centrality in protein networks, Nature 411 (6833) (2001) 41–42.

44. Maxim Kalaev, Mike Smoot, Trey Ideker, Roded Sharan, NetworkBLAST: comparative analysis of protein networks, Bioinformatics 24 (4) (2008) 594–596.

45. Y. Kawahashi, N. Doi, H. Takashima, C. Tsuda, Y. Oishi, R. Oyama, M. Yonezawa, E. Miyamoto-Sato, H. Yanagawa, In vitro protein microarrays for detecting protein-protein interactions: application of a new method for fluorescence labeling of proteins, Proteomics 3 (2003) 1236–1243.

46. Brian P. Kelley, Bingbing Yuan, Fran Lewitter, Roded Sharan, Brent R. Stockwell, Trey Ideker, PathBLAST: a tool for alignment of protein interaction networks, Nucleic Acids Research 32 (suppl_2) (2004) W83–W88.

47. A.D. King, N. Przulj, I. Jurisica, Protein complex prediction via cost-based clustering, Bioinformatics 20 (17) (2004) 3013–3020.

48. A.D. King, Graph clustering with restricted neighbourhood search, Master's thesis, University of Toronto, 2004.

49. Johannes Kobler, Uwe Schöning, Jacobo Torán, The Graph Isomorphism Problem: Its Structural Complexity, Springer Science & Business Media, 2012.
50. Mehmet Koyutürk, Yohan Kim, Umut Topkara, Shankar Subramaniam, Wojciech Szpankowski, Ananth Grama, Pairwise alignment of protein interaction networks, Journal of Computational Biology 13 (2) (2006) 182–199.
51. Oleksii Kuchaiev, Tijana Milenković, Vesna Memišević, Wayne Hayes, Nataša Pržulj, Topological network alignment uncovers biological function and phylogeny, Journal of the Royal Society Interface 7 (50) (2010) 1341–1354.
52. Oleksii Kuchaiev, Nataša Pržulj, Integrative network alignment reveals large regions of global network similarity in yeast and human, Bioinformatics 27 (10) (2011) 1390–1396.
53. Oleksii Kuchaiev, Marija Rašajski, Desmond J. Higham, Nataša Pržulj, Geometric de-noising of protein-protein interaction networks, PLoS Computational Biology 5 (8) (2009) e1000454.
54. Michael Lappe, Liisa Holm, Unraveling protein interaction networks with near-optimal efficiency, Nature Biotechnology 22 (December 2003) 98–103.
55. Chengwei Lei, Jianhua Ruan, A novel link prediction algorithm for reconstructing protein–protein interaction networks by topological similarity, Bioinformatics 29 (3) (2012) 355–364.
56. N. Bertin, S. Li, C.M. Armstrong, et al., A map of the interactome network of the metazoan c. elegans, Science 303 (5657) (2004) 540–543.
57. Chung-Shou Liao, Kanghao Lu, Michael Baym, Rohit Singh, Bonnie Berger, IsorankN: spectral methods for global alignment of multiple protein networks, Bioinformatics 25 (12) (2009) i253–i258.
58. Sridevi Maharaj, Zarin Ohiba, Wayne Hayes, Comparing different graphlet measures for evaluating network model fits to biogrid ppi networks, in: International Conference on Algorithms for Computational Biology, Springer, 2019, pp. 52–67.
59. Noël Malod-Dognin, Kristina Ban, Nataša Pržulj, Unified alignment of protein-protein interaction networks, Scientific Reports 7 (1) (2017) 953.
60. Noël Malod-Dognin, Nataša Pržulj, L-graal: Lagrangian graphlet-based network aligner, Bioinformatics (2015), btv130.
61. Noël Malod-Dognin, Nataša Pržulj, Functional geometry of protein interactomes, Bioinformatics 35 (19) (2019) 3727–3734.
62. Nil Mamano, Wayne Hayes, Sana: simulated annealing network alignment applied to biological networks, arXiv preprint, arXiv:1607.02642, 2016.
63. Vesna Memišević, Nataša Pržulj, C-graal: common-neighbors-based global graph alignment of biological networks, Integrative Biology 4 (7) (2012) 734–743.
64. Lei Meng, Joseph Crawford, Aaron Striegel, Tijana Milenkovic, Igloo: integrating global and local biological network alignment, arXiv preprint, arXiv:1604.06111, 2016.
65. Marianna Milano, Pietro Hiram Guzzi, Mario Cannataro, Glalign: a novel algorithm for local network alignment, in: IEEE/ACM Transactions on Computational Biology and Bioinformatics, 2018.
66. Marianna Milano, Pietro Hiram Guzzi, Olga Tymofieva, Duan Xu, Christofer Hess, Pierangelo Veltri, Mario Cannataro, An extensive assessment of network alignment algorithms for comparison of brain connectomes, BMC Bioinformatics 18 (6) (2017) 235.
67. Tijana Milenković, Weng Leong Ng, Wayne Hayes, Nataša Pržulj, Optimal network alignment with graphlet degree vectors, Cancer Informatics 8 (2010).

68. Marco Mina, Pietro Hiram Guzzi, Improving the robustness of local network alignment: design and extensive assessment of a Markov clustering-based approach, IEEE/ACM Transactions on Computational Biology and Bioinformatics 11 (3) (2014) 561–572.

69. Koyel Mitra, Anne-Ruxandra Carvunis, Sanath Kumar Ramesh, Trey Ideker, Integrative approaches for finding modular structure in biological networks, Nature Reviews. Genetics 14 (10) (2013) 719–732.

70. Yoichi Murakami, Kenji Mizuguchi, Homology-based prediction of interactions between proteins using averaged one-dependence estimators, BMC Bioinformatics 15 (1) (2014) 213.

71. Behnam Neyshabur, Ahmadreza Khadem, Somaye Hashemifar, Seyed Shahriar Arab, Netal: a new graph-based method for global alignment of protein–protein interaction networks, Bioinformatics 29 (13) (2013) 1654–1662.

72. Cédric Notredame, Desmond G. Higgins, Jaap Heringa, T-coffee: a novel method for fast and accurate multiple sequence alignment, Journal of Molecular Biology 302 (1) (2000) 205–217.

73. Roland A. Pache, Arnaud Céol, Patrick Aloy, NetAligner—a network alignment server to compare complexes, pathways and whole interactomes, Nucleic Acids Research 40 (W1) (2012) W157–W161.

74. Lawrence Page, Sergey Brin, Rajeev Motwani, Terry Winograd, The pagerank citation ranking: bringing order to the web, Technical Report 1999-66, Stanford InfoLab, November 1999, Previous number = SIDL-WP-1999-0120.

75. Daniel Park, Rohit Singh, Michael Baym, Chung-Shou Liao, Bonnie Berger, Isobase: a database of functionally related proteins across ppi networks, Nucleic Acids Research 39 (suppl 1) (2011) D295–D300.

76. Clara Pizzuti, Simona E. Rombo, Elena Marchiori, Complex detection in protein-protein interaction networks: a compact overview for researchers and practitioners, in: European Conference on Evolutionary Computation, Machine Learning and Data Mining in Bioinformatics, Springer, 2012, pp. 211–223.

77. N. Przulj, D.G. Corneil, I. Jurisica, Modeling interactome: scale-free or geometric?, Bioinformatics 20 (18) (2004) 3508–3516.

78. N. Przulj, D.A. Wigle, I. Jurisica, Functional topology in a network of protein interactions, Bioinformatics 20 (3) (2004) 340–348.

79. Nataša Pržulj, Biological network comparison using graphlet degree distribution, Bioinformatics 23 (2) (2007) e177–e183.

80. Nataša Pržulj, Protein-protein interactions: making sense of networks via graph-theoretic modeling, BioEssays 33 (2) (2011) 115–123.

81. Teresa M. Przytycka, Yoo-Ah Kim, Network integration meets network dynamics, BMC Biology 8 (1) (2010) 48.

82. J.C. Rain, L. Selig, H. De Reuse, et al., The protein protein interaction map of helicobacter pylori, Nature 409 (6817) (2001) 211–215.

83. G. Rigaut, A. Shevchenko, B. Rutz, M. Wilm, M. Mann, B. Seraphin, A generic protein purification method for protein complex characterization and proteome exploration, Nature Biotechnology 17 (1999) 1030–1032.

84. Swarup Roy, Hazel N. Manners, Ahed Elmsallati, Jugal K. Kalita, Alignment of protein-protein interaction networks, in: Shoba Ranganathan, Michael Gribskov, Kenta Nakai, Christian Schönbach (Eds.), Encyclopedia of Bioinformatics and Computational Biology, Academic Press, Oxford, 2019, pp. 997–1015.

85. Sayed M.E. Sahraeian, Byung-Jun Yoon, Resque: network reduction using semi-Markov random walk scores for efficient querying of biological networks, Bioinformatics 28 (16) (2012) 2129–2136.

86. Sayed Mohammad Ebrahim Sahraeian, Byung-Jun Yoon, Smetana: accurate and scalable algorithm for probabilistic alignment of large-scale biological networks, PLoS ONE 8 (7) (2013) e67995.

87. Lukasz Salwinski, David Eisenberg, Computational methods of analysis of protein–protein interactions, Current Opinion in Structural Biology 13 (3) (2003) 377–382.

88. Vikram Saraph, Tijana Milenković, Magna: maximizing accuracy in global network alignment, Bioinformatics 30 (20) (2014) 2931–2940.

89. Roded Sharan, Silpa Suthram, Ryan M. Kelley, Tanja Kuhn, Scott McCuine, Peter Uetz, Taylor Sittler, Richard M. Karp, Trey Ideker, Conserved patterns of protein interaction in multiple species, Proceedings of the National Academy of Sciences of the United States of America 102 (6) (2005) 1974–1979.

90. Yu-Keng Shih, Srinivasan Parthasarathy, Scalable global alignment for multiple biological networks, BMC Bioinformatics 13 (Suppl 3) (2012) S11.

91. Rohit Singh, Daniel Park, Jinbo Xu, Raghavendra Hosur, Bonnie Berger, Struct2net: a web service to predict protein–protein interactions using a structure-based approach, Nucleic Acids Research 38 (suppl 2) (2010) W508–W515.

92. Rohit Singh, Jinbo Xu, Bonnie Berger, Global alignment of multiple protein interaction networks with application to functional orthology detection, Proceedings of the National Academy of Sciences 105 (35) (2008) 12763–12768.

93. E.A. Smith, R.M. Corn, Surface plasmon resonance imaging as a tool to monitor biomolecular interactions in an array based format, Applied Spectroscopy 57 (2003) 320A–332A.

94. Michael P.H. Stumpf, Thomas Thorne, Eric de Silva, Ronald Stewart, Hyeong Jun An, Michael Lappe, Carsten Wiuf, Estimating the size of the human interactome, Proceedings of the National Academy of Sciences 105 (19) (2008) 6959–6964.

95. Panagiotis Symeonidis, Nantia Iakovidou, Nikolaos Mantas, Yannis Manolopoulos, From biological to social networks: link prediction based on multi-way spectral clustering, Data & Knowledge Engineering 87 (2013) 226–242.

96. Roman L. Tatusov, Eugene V. Koonin, David J. Lipman, A genomic perspective on protein families, Science 278 (5338) (1997) 631–637.

97. Hanghang Tong, Christos Faloutsos, Jia-Yu Pan, Random walk with restart: fast solutions and applications, Knowledge and Information Systems 14 (3) (2008) 327–346.

98. Peter Uetz, Loic Giot, Gerard Cagney, Traci A. Mansfield, Richard S. Judson, James R. Knight, Daniel Lockshon, Vaibhav Narayan, Maithreyan Srinivasan, Pascale Pochart, et al., A comprehensive analysis of protein–protein interactions in saccharomyces cerevisiae, Nature 403 (6770) (2000) 623–627.

99. S. van Dongen, Graph Clustering by Flow Simulation, PhD thesis, University of Utrecht, May 2000.

100. V. Vijayan, Vikram Saraph, T. Milenković, Magna++: maximizing accuracy in global network alignment via both node and edge conservation, Bioinformatics (2015), btv161.

101. C. von Mering, R. Krause, B. Snel, M. Cornell, S.G. Oliver, S. anf Bork, P. Fields, Comparative assessment of large-scale data sets of protein–protein interactions, Nature 417 (2002) 399–403.

102. Hua Wang, Heng Huang, Chris Ding, Feiping Nie, Predicting protein–protein interactions from multimodal biological data sources via

nonnegative matrix tri-factorization, Journal of Computational Biology 20 (4) (2013) 344–358.

103. Yoshihiro Yamanishi, J-P. Vert, Minoru Kanehisa, Protein network inference from multiple genomic data: a supervised approach, Bioinformatics 20 (suppl 1) (2004) i363–i370.

104. Qiangfeng Cliff Zhang, Donald Petrey, Lei Deng, Li Qiang, Yu Shi, Chan Aye Thu, Brygida Bisikirska, Celine Lefebvre, Domenico Accili, Tony Hunter, et al., Structure-based prediction of protein–protein interactions on a genome-wide scale, Nature 490 (7421) (2012) 556.

105. H. Zhu, M. Bilgin, D. Bangham, A. Hall, P. Casamayor, N. Bertone, R. Lan, S. Jansen, T. Bidlingmaier, T. Houfek, P. Mitchell, R.A. Miller, M. Dean, M. Gerstein, M. Snyder, Global analysis of protein activities using proteome chips, Science 293 (2001) 2101–2105.

106. Elena Zotenko, Katia S. Guimarães, Raja Jothi, Teresa M. Przytyc, Decomposition of overlapping protein complexes: a graph theoretical method for analyzing static and dynamic protein associations, Algorithms for Molecular Biology 1 (7) (2006) 1–7.

Brain connectome networks and analysis

Contents

The human brain is a complex system, whose organization has been represented as a structural connectome of interconnected areas and a functional connectome of interregional neural activity. The human brain connectome can be mapped using MRI data, and further characterized through strategies based on graph theory. The modeling and analysis of connectome is a big challenge in neurosciences and network research. This chapter highlights main techniques for acquiring the process of brain networks, graph theoretic modeling, and analysis of connectome networks.

8.1 Human brain connectome

The human brain known to be one of the most complex interconnected systems consists of 80 to 100 billion nerve cells or neurons. Neurons are connected by synapses and an estimated 100 trillion connections between them. Neurons are organized into regions given by the connections. Connections may be analyzed from an anatomical point of view, i.e., the set of physical connections among neurons, and from a functional point of view, i.e., functional regions composed of groups of neurons that perform different functions. Variations of the set of connections are strictly linked to the insurgency and progression of neurological diseases as demonstrated by many works [4,9,13,29]. From a neuroscience point of view, it is always interesting to elucidate how the brain actually work with the help of such a large interconnections of nerve cells. Consequently, the modeling of the whole system of the brain elements and their relations has become an emerging theme in neuroscience. Initially, it has been assumed that a particular function of a brain is mapped to a particular region of the brain consisting of millions of neurons. Recent studies, however, discovered certain new facts that such functional regions do not work in isolation, rather, they interact with each other to perform mental activities. Main pillars of such research area are the availability of novel technologies that able to investigate such connections and to model such data using graph theory.

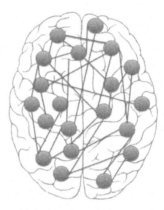

Figure 8.1. An illustrative view of brain network. The nodes represent active regions of the brain and edge signifies possible physical or functional links between the regions.

The development of novel technologies for imaging the brain has given the researchers the possibility to deeply study the anatomical and functional organization of the brain. This caused the development of a new research field, named *connectomics*, that focuses on the organization of the connectome, i.e., the whole set of associations (both physical and functional) of the constitutive elements of the brain [7,10,28,42]. With the availability of high-resolution neuroimaging technologies, such as fMRI and diffusion spectrum imaging (DSI), it is now possible to study brain activities *in silico* and noninvasive way. Modern neuroimaging techniques open up new avenues to study the brain and its activities through the light of how complex interplay happens between the various neural cells. Hence, it is obvious to view the mechanism of interactions as network of neural cells.

The set of connections among neural cells inside the brain is termed as *connectome*. Connectome is the comprehensive wiring architecture of neural connection of brain. The production and study of connectomes are referred to as connectomics. Study of connectome having an increasingly prominent role in bioinformatics and neurosciences in general. The key assumption of this field is the modeling of the brain as a complex (heterogeneous) network based on data extracted from neuroimaging. Consequently, the analysis of the representation of such a graph is giving valuable knowledge to the researchers. Graph theory models the human brain as a network of nodes linked by edges. The nodes represent brain regions, whereas the edges represent fibre connections in structural data and temporal correlations in functional data. A mapping of brain image to network can be viewed illustratively as shown in Fig. 8.1.

To construct a connectome network from the neuroimaging data, one needs to define first the region of interests (nodes) and association among the nodes (edges). However, the definition of nodes and edges to form a connectome network depends highly on the imaging inputs used. The networks may be either structural or functional, representing static or dynamic behavior of the brain cell activities, respectively.

High-resolution MRI images are used as a starting point for deriving brain graphs. Starting from neuroimages and other potential sources of brain activity data, such as electroencephalogram (EEG) or magnetoencephalogram (MEG), the parcelation process identifies brain regions and assigns each region to a node. The parcelation-based brain graph is then derived. Once the network or graph is generated, a set of graph analysis algorithms are applied to analyze the network to identify hub nodes (hub analysis) [46], or communities representing functionally related nodes [11,33]. Fig. 8.2 shows a typical workflow involved in connectome network analysis.

We next discuss the methodologies for capturing the network from the neuroimaging data.

8.2 From neuroimaging to brain graph

The process of building a network, describing the brain, starts with the acquiring images from the brain. Images are then processed for the construction of the network. Next, an adjacency matrix is created that highlights the connectivity between the regions.

Differently from other application fields, e.g., protein interaction networks, some challenges are still open. One of the main challenges is the absence of a commonly accepted definition of

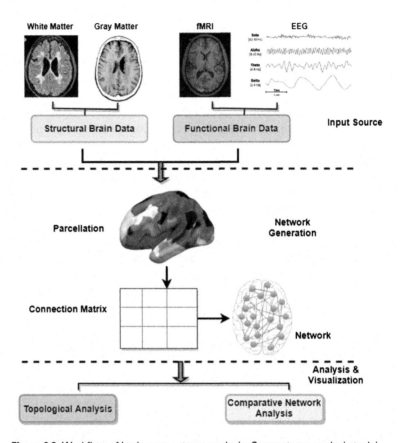

Figure 8.2. Workflow of brain connectome analysis. Connectome analysis task is divided into three major activities: image/input acquisition, network construction through parcellation and post inference analysis and visualization.

a node in a brain network [42]. Therefore there exist two main approaches: structural connectomics (which uses single neurons as nodes) and functional connectomics (uses functional regions as nodes). There exist different imaging techniques that may be used to generate graphs. Interested reader may find some more details in [28].

In the following, we present the noninvasive neuroimaging solutions to yield structural and functional connectivity.

The workflow, as depicted in Fig. 8.2, starts with magnetic resonance imaging (MRI). The set of scans acquired in a single session is then used to register the locations of the brain with respect to a set of known regions. Functional connectivity is established using a time series of acquisitions and the analysis of corresponding voxels. The correlation between the time series of different voxels or, using aggregated measures, brain regions can be detected and represented as a correlation matrix (value ranging from -1 to 1). Connections among regions are then derived from the correlation matrix that may be interpreted as a weighted or a binary network after thresholding all the values.

On the other hand, structural connectivity is established by acquiring diffusion tensor imaging (DTI) or DSI. The post-processing of these images with tracking algorithms (both deterministic and probabilistic) yield to the individuation of a set of streamlines between brain regions. Once again the streamlines may be interpreted as a connection matrix, in which each element represents the connection (or its probability) among two brain regions.

An essential step in the analysis of both images is the subdivision of the brain into regions, also referred to as a **parcelation process**. Formally the brain parcelation is the process of partitioning the brain into a set of regions that are homogeneous (from a biological point of view) and non-overlapping.

Despite the relevance of this process, it should be noted that there is not an universally accepted definition of parcelation of the brain into homogeneous regions [28], as well as of the connections among them. Consequently, in connectomics there is no clear definition of nodes and edges. Therefore there exist many different parcelation methods that produces graphs with different properties.

- Atlas-Based parcelation: This is the process of parcelation of the brain by using predefined anatomical templates also known as atlas. In this approach, structural images from MRI are registered with respect to anatomical atlas based on the template defined by Brodmann areas. The whole brain is subdivided into labeled regions according to the different labeled regions of the templates. This process may have some pitfalls when analyzing brains that differ in a significant way from atlas, e.g., in the case of children's or neurological diseases.
- Random template-based parcelation: This is the process of using randomly generated templates. Many different algorithms are used to produce parcels that have approximately the same size in order to produce regions with similar size [18].
- Connectivity-based parcelation: It relies on the process to evidence brain regions by investigating patterns of connections (either structural or functional) under the hypothesis that regions that present a similar connectivity pattern have the same role. Usually algorithms in this class first find small seed regions, then they merge together similar regions to find regions corresponding to brain areas [28].

Each method presents both advantages and pitfalls. For instance, the atlas-based parcelation, i.e., the registration of the brain of the studied patient to the Brodmann template suffers of low accuracy.

This approach is limited by intersubject variability and can be especially problematic in the context of brain maturation.

8.3 Data storage and querying

Data related to brain networks are both images and derived graphs. Images, e.g., magnetic resonance images, are stored into many databases that the user can download for building the brain graph. Here, we report only the databases storing structural or functional matrix for analysis. The brain connectivity toolbox (BCT)[1] [36] contains a collection of large-scale connectivity data sets. The dataset is freely available for download.

The UCLA multimodal connectivity database (UMCD)[2] is an open website for brain network analysis and a repository for connectivity matrices derived from research studies. Network repository (NR)[3] [35] is an interactive data for managing of network data using the web. For more details one can refer Chapter 5.

8.4 Analysis of brain networks: tools and methodologies

In this section, we present the analysis that can be performed on brain graph by distinguishing among the class of measures that quantify the properties of a single graph and the techniques for performing a comparison between networks.

8.4.1 Topological characterization of connectome networks

Once a brain graph is built, there is the need to characterize the obtained graph by measuring its relevant topological properties [26]. Different metrics from statistics, physics and the area of complex networks are widely used in the study of human brain networks. The main measures are the measures of influence of segregation and of integration [39]. Measures of influence aim to

---

[1]https://sites.google.com/site/bctnet/.
[2]http://umcd.humanconnectomeproject.org.
[3]http://networkrepository.com/.

quantify the relevance (or the role) of a node in a network. In connectomics, most used measures are classical centrality measures.

Degree distribution and hub analysis

Brain network analysis aims to detect pivotal regions and their connections and the flow of information among them that enable the functions of the brain. In graph theory pivotal regions are usually defined as *hub nodes* and connections as *bridge edges*. Many studies, lesions on human brain, have suggested that hub and bridges are related to vital neurocognitive functions, and on the spreads of disease in clinical brain disorders [24]. It has been evinced that the loss of hubs or bridges could reduce the effective information flow through the brain network. These properties have been described even in several other mammalian species. The identification of hubs and bridges resides in the definition of *central nodes* through centrality measures.

Literature contains many studies that identify hubs and bridges using centrality measures. Here, we briefly introduce the biological meaning of such measures, and the interested reader may find many details in [47].

One of the first topological measures that has been used is the degree centrality that is defined as the number of edges that are bound to a given node. Therefore a high number of edges suggest an important role of the node. Such measure represent a local measure of centrality (i.e., the centrality of a node with respect to its neighbors). The degree centrality (CD) is defined as the number of edges connected to a node; it is used to quantify the local centrality of each node. Degree centrality is often analyzed for all the nodes. The degrees of all nodes in the network comprise the degree distribution, which is an important marker of network development and tolerance to faults [24].

First studies focused on the determination of the degree distribution for brain networks and on the study of the possible changes of such distribution, considering different species, diseases, and also spatial and temporal scales. Such studies demonstrated the existence of scale-free organization for some species [39]. Power law structures are characterized by the presence of highly connected nodes, i.e., nodes that have a high degree. In parallel, it should be noted that some studies provided the evidence of an exponentially truncated power law, i.e., a network with many hub than a random network, but with few edges than a classical scale-free network. These characteristics have been evidenced in human and cats using functional imaging and region level analysis, whereas neuron-level analysis evidenced the existence of a power-law distribution [22,32]. There is now strong evidence that human brain networks generally have small-world properties of high clustering and high global efficiency [1], a modular community structure [14,27], and heavy-tailed degree distributions [14] that indicate a number of highly connected nodes or hubs [41].

Rich-club organization

Different studies also demonstrated the presence of the *rich-club* phenomenon. A rich-club organization of a network is present when hubs are densely connected and loosely connected with nodes of a lower degree (see previous section for a more deep discussion). The name "rich-club" was introduced in the study of social systems, in which highly important (or central) people, e.g., rich people, form a highly interconnected club [12]. The presence of the so-called rich-club organization can provide insights about the resilience and specialization of the network. For instance, the absence of rich-club organization in protein interaction networks may be related to the high level of functional specialization of proteins, therefore the presence may be related to the absence of high specialization.

It has been claimed that the rich-club within a brain may play a role of a communication backbone of the brain [48]. A study on Schizophrenia affected brain reveals a significant reduction in global connectivity and the strength of hubs, especially in frontal and parietal regions [15]. In another study, reduction in the connectivity of rich-club regions is also observed, leading to lowering in the communication capacity of core brain, effecting the strength of global information processing [48].

Flow of information: segregation and integration

Essential concepts in brain networks are segregation and integration [38,43]. Segregation refers to the modularity of the network, i.e., the propensity of a network to organize its nodes into cliques or communities. In a structural network, the presence of clusters is indicative of functional segregation, that is, the occurrence of processing within densely interconnected groups of brain regions. In functional networks, the presence of clusters indicates statistical dependencies during neural processing.

An important measure of segregation is the *clustering coefficient* [49], which measures the density of connections among nodes and their neighbors. If these neighbors are densely interconnected, they can form a cluster or clique, and they are likely to share specific information.

A variant of the clustering coefficient is the *transitivity* [30]. However, the transitivity is not defined for individual nodes yet. Another standard measure of segregation is the *modularity* [31],

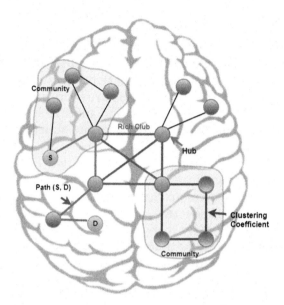

Figure 8.3. Various topological measures used in a brain network analysis such as community, rich club organization, hub nodes, path and clustering coefficient.

which indicates the capability of a network to be subdivided into delineated and non-overlapping groups.

Integration refers to the capacity of the network to being interconnected and to exchange information from distributed brain regions. Measures of integration are based on the concept of paths and path lengths. A path is a sequence of edges that connect two nodes with another one, and its length consists of the distance to reach it. The lengths of paths are indicative of functional integration, that is, the brain's ability to combine specialized information among the regions quickly. Shorter paths imply stronger integration. The most commonly used measure of integration is the *characteristic path length* [49], the average shortest path length between all pairs of nodes in the network. An illustrative brain network with various topological parameters are shown in Fig. 8.3.

Different studies [20,25,50] conducted on Alzheimer's disease have uncovered altered small-world organization due to increased characteristic path length and clustering coefficient. The increase of path length is interpreted as a loss of efficiency of communication between distant brain regions, whereas the increase of clustering coefficient is interpreted as a stronger local specialization. Furthermore, reductions of number of hubs and decreased betweenness centrality and local efficiency caused by β-amyloid accumulation are also detected [3,8,20,37,50].

Community detection in brain graphs

Community detection algorithms are useful to study the organization of brain networks [2], since they provide information about the modular organization of the networks and the presence of network hierarchies. Many studies have been applied to structural and functional brain networks and, more recently, to multimodal networks (i.e., networks that integrate both structural and functional aspects).

Communities are related to the distribution of the degrees [17] that presents significant nonhomogeneities, giving rise to the presence of some vertices that have many edges and few connecting other vertices. Therefore, a community may be defined as a group of nodes that have many edges and few edges connecting them to the rest of the graph. The presence of communities may be seen as indicative of a modular organization in a network. For instance, in PPI networks, protein complexes are evidenced by small dense communities.

In brain graphs, many works presented different definitions of communities, based on slightly different definition of *modularity*. Briefly, many authors tried to adapt the modularity definition to better elucidate both functional and structural networks, and to take into account differences among individuals and the impact of ageing and neurological diseases [6,40]. Therefore there is no complete agreement on the existence of a modularity structure, since the conclusion of the existing works is strictly related to the input data.

As evidence in a recent work [6], it should be noted that network organization of the human brain shows significant variations among individuals, and it is dependent on age and disease status. Therefore the main challenge of community discovery methods for neurosciences is the development of a methodological framework for evidencing modular structures and for discriminating healthy/diseased status for the purposes of multi-scale analysis and biomarker generation.

A recent study on athletes' brain networks [19] reveals a novel interpretation of core-periphery and rich club organization of brain graph. It reports an unification of modularity and core periphery structure. Authors presented a novel approach based on the weighted stochastic block model, demonstrating that functional brain networks show rich mesoscale organization beyond that sought by modularity maximization techniques. The organization is also affected by ageing and by neurological disease. In addition, individual differences are also appreciable.

8.4.2 Comparison of brain networks

Comparison of brain networks is useful for exploring connectivity relationships in individual subjects or between groups of subjects. In individual subjects, comparisons can detect information on structural and functional connectivity relationships [21]. Across groups of subjects, such as healthy vs patients, the comparisons can uncover the abnormalities of network connectivity in different disorders [5,36].

Comparison is made looking at the variations of quantitative measures describing network features we presented in previous sections. This comparison is based on the assumption that compared brains have the same dimensions, and then the related graphs have the same number of nodes. Any difference in size may produce distortions to identify topological differences. Furthermore, the analysis can be a focus on subnetworks of graphs to detect subgraphs discriminating two brain graphs [9,34,51].

However, the use of single global network measures may lack the proper comparison of the substructures. Consequently, identifying an accurate node mapping between networks may offer more details on the comparison, [23,44].

8.4.3 Analyzing brain network in R: the BrainGraph package

The interest for analyzing brain data has lead to the introduction of many packages for brain data analysis, spanning different subfields: from image to graph analysis. Literature contains both specialized packages and packages that preprocess image data, and then use existing algorithms and packages for graph data analysis. For instance the `BrainGraph` package [45] implemented in R language provides a set of tools for performing analysis on brain graphs. It is an extension of `igraph` package, as discussed earlier.

It is able to import data produced by Freesurfer [16] software suite (e.g., brain volumes, surface area), diffusion tensor tractography data produced by FSL.[4] A typical flow of analysis of brain data in BrainGraph may be subdivided into following steps:

Step 1: Import data   First users need to import data to create connectivity matrices (e.g., adjacency matrices of the graph). This step is strictly related to the kind of input data. For instance, for cortical thickness or partial correlations, data user needs to preprocess data to derive matrices from these data. For DTI tractography or resting state fMRI, the user may directly import the connectivity matrices produced from the neuroimaging software.

[4]http://www.fmrib.ox.ac.uk/fsl.

> ### Importing data into BrainGraph from DPABI software
>
> ```
> files <- list.files(list.dirs(datadir, recursive=T),'
>     ROICorrelation_[a–z]+.*.txt', full.names=T)
> ```

Step 2: Thresholding matrices   After the data has been loaded, the user needs to threshold matrices to create the correct adjacency matrices. The choice of the thresholding method is related to the kind of input data. Braingraph package suggests the right choice for each possible input.

Step 3: Graph analysis   At this step the user has created the graph(s) that have to be analyzed; he may use the in-built functions to analyze the graph data, e.g., for testing the small-world or rich-club properties.

The usual way to test these properties is to generate a set of random graph having same properties of the analyzed one(s)—representing null model—then to calculate the parameter for each graph and finally to evaluate the statistical difference of the parameters of the analyzed graphs with respect to the null model. The following fragment of code generates 100 random graphs, then calculates the required properties:

> ### Testing small-world and rich-club properties
>
> ```
> NumofRandGraph <- 100 \\
> random_vars <- analysis_random_graphs(g, kNumRand,
>     savedir=outdir, clustering=clustering)11
> rich.dt <- random_vars$rich
> rich.dt <- rich.dt[complete.cases(rich.dt)]
> NAsmall.dt <- random_vars$small
> rand.dt <- random_vars$rand
> ```

8.5 Summary

Here, we presented a review on computational aspects of the brain network. We tried to elucidate some of the conceptual aspects and methodologies that concern (i) the building of a brain graph, (ii) the measures that may be applied to capture the organization of brain networks, (iii) the comparison of the brain networks, (iv) the tools for the network analysis of the brain, and (v) the databases publicly available for storing the brain networks.

Acknowledgment

Authors thank Dr. Marianna Milano and Prof. Mario Cannataro for their suggestions and collaboration while writing this chapter.

References

1. Sophie Achard, Ed Bullmore, Efficiency and cost of economical brain functional networks, PLoS Computational Biology 3 (2) (2007) e17.
2. Arian Ashourvan, Qawi K. Telesford, Timothy Verstynen, Jean M. Vettel, Danielle S. Bassett, Multi-scale detection of hierarchical community architecture in structural and functional brain networks, PLoS ONE 14 (5) (2019) e0215520.
3. Feng Bai, Ni Shu, Yonggui Yuan, Yongmei Shi, Hui Yu, Di Wu, Jinhui Wang, Mingrui Xia, Yong He, Zhijun Zhang, Topologically convergent and divergent structural connectivity patterns between patients with remitted geriatric depression and amnestic mild cognitive impairment, The Journal of Neuroscience 32 (12) (2012) 4307–4318.
4. Cornelia I. Bargmann, Eve Marder, From the connectome to brain function, Nature Methods 10 (6) (2013) 483–490.
5. Danielle S. Bassett, Edward Bullmore, Beth A. Verchinski, Venkata S. Mattay, Daniel R. Weinberger, Andreas Meyer-Lindenberg, Hierarchical organization of human cortical networks in health and schizophrenia, The Journal of Neuroscience 28 (37) (2008) 9239–9248.
6. Richard F. Betzel, Alessandra Griffa, Andrea Avena-Koenigsberger, Joaquín Goñi, Jean-Philippe Thiran, Patric Hagmann, Olaf Sporns, Multi-scale community organization of the human structural connectome and its relationship with resting-state functional connectivity, Network Science 1 (3) (2013) 353–373.
7. Bharat B. Biswal, Maarten Mennes, Xi-Nian Zuo, Suril Gohel, Clare Kelly, Steve M. Smith, Christian F. Beckmann, Jonathan S. Adelstein, Randy L. Buckner, Stan Colcombe, et al., Toward discovery science of human brain function, Proceedings of the National Academy of Sciences 107 (10) (2010) 4734–4739.
8. Randy L. Buckner, Jorge Sepulcre, Tanveer Talukdar, Fenna M. Krienen, Hesheng Liu, Trey Hedden, Jessica R. Andrews-Hanna, Reisa A. Sperling, Keith A. Johnson, Cortical hubs revealed by intrinsic functional connectivity: mapping, assessment of stability, and relation to Alzheimer's disease, The Journal of Neuroscience 29 (6) (2009) 1860–1873.
9. Edward T. Bullmore, Danielle S. Bassett, Brain graphs: graphical models of the human brain connectome, Annual Review of Clinical Psychology 7 (2011) 113–140.
10. Mario Cannataro, Pietro Hiram Guzzi, Alessia Sarica, Data mining and life sciences applications on the grid, Wiley Interdisciplinary Reviews: Data Mining and Knowledge Discovery 3 (3) (2013) 216–238.
11. Michael W. Cole, Danielle S. Bassett, Jonathan D. Power, Todd S. Braver, Steven E. Petersen, Intrinsic and task-evoked network architectures of the human brain, Neuron 83 (1) (2014) 238–251.
12. Vittoria Colizza, Alessandro Flammini, M. Angeles Serrano, Alessandro Vespignani, Detecting rich-club ordering in complex networks, Nature Physics 2 (2) (2006) 110.

13. Eric Courchesne, Karen Pierce, Why the frontal cortex in autism might be talking only to itself: local over-connectivity but long-distance disconnection, Current Opinion in Neurobiology 15 (2) (2005) 225–230.

14. M. Victor Eguiluz, Dante R. Chialvo, Guillermo A. Cecchi, Marwan Baliki, A. Vania Apkarian, Scale-free brain functional networks, Physical Review Letters 94 (1) (2005) 018102.

15. Massimo Filippi, Martijn P. van den Heuvel, Alexander Fornito, Yong He, Hilleke E. Hulshoff Pol, Federica Agosta, Giancarlo Comi, Maria A. Rocca, Assessment of system dysfunction in the brain through mri-based connectomics, The Lancet Neurology 12 (12) (2013) 1189–1199.

16. Bruce Fischl, Freesurfer, NeuroImage 62 (2) (2012) 774–781.

17. Santo Fortunato, Community detection in graphs, Physics Reports 486 (3–5) (2010) 75–174.

18. Stefan Geyer, Marcel Weiss, Katja Reimann, Gabriele Lohmann, Robert Turner, Microstructural parcellation of the human cerebral cortex–from brodmanns post-mortem map to in vivo mapping with high-field magnetic resonance imaging, Frontiers in Human Neuroscience 5 (2011) 19.

19. Shi Gu, Cedric Huchuan Xia, Rastko Ciric, Tyler M. Moore, Ruben C. Gur, Raquel E. Gur, Theodore D. Satterthwaite, Danielle S. Bassett, Unifying modular and core-periphery structure in functional brain networks over development, arXiv preprint, arXiv:1904.00232, 2019.

20. Yong He, Zhang Chen, Alan Evans, Structural insights into aberrant topological patterns of large-scale cortical networks in Alzheimers disease, The Journal of Neuroscience 28 (18) (2008) 4756–4766.

21. Christopher J. Honey, Rolf Kötter, Michael Breakspear, Olaf Sporns, Network structure of cerebral cortex shapes functional connectivity on multiple time scales, Proceedings of the National Academy of Sciences 104 (24) (2007) 10240–10245.

22. Yasser Iturria-Medina, Roberto C. Sotero, Erick J. Canales-Rodríguez, Yasser Alemán-Gómez, Lester Melie-García, Studying the human brain anatomical network via diffusion-weighted mri and graph theory, NeuroImage 40 (3) (2008) 1064–1076.

23. Sofia Ira Ktena, Salim Arslan, Sarah Parisot, Daniel Rueckert, Exploring heritability of functional brain networks with inexact graph matching, arXiv preprint, arXiv:1703.10062, 2017.

24. Hunki Kwon, Yong-Ho Choi, Jong-Min Lee, A physarum centrality measure of the human brain network, Scientific Reports 9 (1) (2019) 5907.

25. Chun-Yi Lo, Pei-Ning Wang, Kun-Hsien Chou, Jinhui Wang, Yong He, Ching-Po Lin, Diffusion tensor tractography reveals abnormal topological organization in structural cortical networks in Alzheimers disease, The Journal of Neuroscience 30 (50) (2010) 16876–16885.

26. Djalel Eddine Meskaldji, Elda Fischi-Gomez, Alessandra Griffa, Patric Hagmann, Stephan Morgenthaler, Jean-Philippe Thiran, Comparing connectomes across subjects and populations at different scales, NeuroImage 80 (2013) 416–425.

27. David Meunier, Sophie Achard, Alexa Morcom, Ed Bullmore, Age-related changes in modular organization of human brain functional networks, NeuroImage 44 (3) (2009) 715–723.

28. Marianna Milano, Pietro Hiram Guzzi, Mario Cannataro, Network building and analysis in connectomics studies: a review of algorithms, databases and technologies, Network Modeling Analysis in Health Informatics and Bioinformatics 8 (1) (2019) 13.

29. Marianna Milano, Pietro Hiram Guzzi, Mario Cannataro, Using multiple network alignment for studying connectomes, Network Modeling Analysis in Health Informatics and Bioinformatics 8 (1) (2019) 5.

30. Mark E.J. Newman, The structure and function of complex networks, SIAM Review 45 (2) (2003) 167–256.
31. Mark E.J. Newman, Michelle Girvan, Finding and evaluating community structure in networks, Physical Review E 69 (2) (2004) 026113.
32. Josef Parvizi, Gary W. Van Hoesen, Joseph Buckwalter, Antonio Damasio, Neural connections of the posteromedial cortex in the macaque, Proceedings of the National Academy of Sciences 103 (5) (2006) 1563–1568.
33. Jonathan D. Power, Alexander L. Cohen, Steven M. Nelson, Gagan S. Wig, Kelly Anne Barnes, Jessica A. Church, Alecia C. Vogel, Timothy O. Laumann, Fran M. Miezin, Bradley L. Schlaggar, et al., Functional network organization of the human brain, Neuron 72 (4) (2011) 665–678.
34. Jonas Richiardi, Hamdi Eryilmaz, Sophie Schwartz, Patrik Vuilleumier, Dimitri Van De Ville, Decoding brain states from fmri connectivity graphs, NeuroImage 56 (2) (2011) 616–626.
35. Ryan A. Rossi, Nesreen K. Ahmed, Networkrepository: a graph data repository with visual interactive analytics, arXiv preprint, arXiv:1410.3560, 2014.
36. Mikail Rubinov, Olaf Sporns, Complex network measures of brain connectivity: uses and interpretations, NeuroImage 52 (3) (2010) 1059–1069.
37. Ni Shu, Ying Liang, He Li, Junying Zhang, Xin Li, Liang Wang, Yong He, Yongyan Wang, Zhanjun Zhang, Disrupted topological organization in white matter structural networks in amnestic mild cognitive impairment: relationship to subtype, Radiology 265 (2) (2012) 518–527.
38. Olaf Sporns, Network attributes for segregation and integration in the human brain, Current Opinion in Neurobiology 23 (2) (2013) 162–171.
39. Olaf Sporns, Structure and function of complex brain networks, Dialogues Clinical Neuroscience 15 (3) (2013) 247–262.
40. Olaf Sporns, Richard F. Betzel, Modular brain networks, Annual Review of Psychology 67 (2016) 613–640.
41. Olaf Sporns, Christopher J. Honey, Rolf Kötter, Identification and classification of hubs in brain networks, PLoS ONE 2 (10) (2007) e1049.
42. Olaf Sporns, Giulio Tononi, Rolf Kotter, The human connectome: a structural description of the human brain, PLoS Computational Biology 1 (4) (2005) e42.
43. Cornelis J. Stam, Jaap C. Reijneveld, Graph theoretical analysis of complex networks in the brain, Nonlinear Biomedical Physics 1 (1) (2007) 3.
44. Olga Tymofiyeva, Etay Ziv, A. James Barkovich, Christopher P. Hess, Duan Xu, Brain without anatomy: construction and comparison of fully network-driven structural mri connectomes, PLoS ONE 9 (5) (2014) e96196.
45. Nathalie Tzourio-Mazoyer, Brigitte Landeau, Dimitri Papathanassiou, Fabrice Crivello, Olivier Etard, Nicolas Delcroix, Bernard Mazoyer, Marc Joliot, Automated anatomical labeling of activations in spm using a macroscopic anatomical parcellation of the mni mri single-subject brain, NeuroImage 15 (1) (2002) 273–289.
46. Martijn P. van den Heuvel, René S. Kahn, Joaquín Goñi, Olaf Sporns, High-cost, high-capacity backbone for global brain communication, Proceedings of the National Academy of Sciences 109 (28) (2012) 11372–11377.
47. Martijn P. Den Van Heuvel, Olaf Sporns, Rich-club organization of the human connectome, The Journal of Neuroscience 31 (44) (2011) 15775–15786.
48. Martijn P. van den Heuvel, Olaf Sporns, Guusje Collin, Thomas Scheewe, René C.W. Mandl, Wiepke Cahn, Joaquín Goñi, Hilleke E. Hulshoff Pol, René S. Kahn, Abnormal rich club organization and functional brain dynamics in schizophrenia, JAMA Psychiatry 70 (8) (2013) 783–792.

49. Duncan J. Watts, Steven H. Strogatz, Collective dynamics of small-world networks, Nature 393 (6684) (1998) 440–442.

50. Zhijun Yao, Yuanchao Zhang, Lei Lin, Yuan Zhou, Cunlu Xu, Tianzi Jiang, Alzheimers disease neuroimaging initiative et al. abnormal cortical networks in mild cognitive impairment and Alzheimers disease, PLoS Computational Biology 6 (11) (2010) e1001006.

51. Andrew Zalesky, Alex Fornito, Ian H. Harding, Luca Cocchi, Murat Yücel, Christos Pantelis, Edward T. Bullmore, Whole-brain anatomical networks: does the choice of nodes matter?, NeuroImage 50 (3) (2010) 970–983.

Conclusion

This book presented an overview on concepts and research related to biological networks, focusing on three major application fields: gene expression networks, protein interaction networks, and brain connectome networks. We tried to make the book interesting for the early and middle stage researchers who started researches in analyzing and inferring any three of the networks.

We are confident about its resourcefulness due every chapter's contents. We even briefly introduced the basic graph theory and its analysis, keeping in mind any noncomputer science readers, interested in network researches. R is a powerful, rich, and freely available scripting tool that can be used for network analysis. We demonstrated how R scripts can be used for graph analysis. We briefly discussed various algorithms. The interested reader may find useful resources (websites, data repository, and original papers) in references provided for each chapter.

Biological network inference and analysis is still growing, and it requires rigorous improvement of the methods for achieving qualitative results. It is always not straightforward to describe any biological molecular interaction mechanism with the help of simplistic mathematical models. Usually interaction mechanisms are nonlinear in nature, which makes the task of biological network inference and analysis more challenging. Sometime unavailability or lack of adequate input data sources add to the challenging nature of the task more. Due to the lack of gold standard networks, it is even difficult to validate any newly proposed regulatory network inference methods to justify its credibility. Combination of multisource and multitype data inputs may improve the outcome of any inference and analysis method. However, computationally it is an expensive task. The situation becomes worst when we need to deal with dynamic networks of large size available in nonuniform formats. Hence, the overall problem of biological network inference and analysis mapped to the problem of big-data analysis. An effective big-data analytics platform is a present need for handling huge, incremental network analysis. Accordingly, powerful and suitable parallel platform design is important for the same. Existing algorithms are limited in handling few hundreds nodes and edges in a network. To make them suitable for

Biological Network Analysis. https://doi.org/10.1016/B978-0-12-819350-1.00015-3

parallel environments requires complete redesigning of the algorithms. Moreover, the rising of novel methods for network embeddings may also produces dramatic changes in network analysis.

Index

Printed in the United States
By Bookmasters